12/13

2

HAZARDOUS WASTES, INDUSTRIAL DISASTERS, AND ENVIRONMENTAL HEALTH RISKS

HAZARDOUS WASTES, INDUSTRIAL DISASTERS, AND ENVIRONMENTAL HEALTH RISKS

LOCAL AND GLOBAL ENVIRONMENTAL STRUGGLES

Francis O. Adeola

palgrave
macmillan

HAZARDOUS WASTES, INDUSTRIAL DISASTERS, AND ENVIRONMENTAL HEALTH RISKS
Copyright © Francis O. Adeola, 2011.

All rights reserved.

First published in 2011 by
PALGRAVE MACMILLAN®
in the United States—a division of St. Martin's Press LLC,
175 Fifth Avenue, New York, NY 10010.

Where this book is distributed in the UK, Europe and the rest of the world,
this is by Palgrave Macmillan, a division of Macmillan Publishers Limited,
registered in England, company number 785998, of Houndmills,
Basingstoke, Hampshire RG21 6XS.

Palgrave Macmillan is the global academic imprint of the above companies
and has companies and representatives throughout the world.

Palgrave® and Macmillan® are registered trademarks in the United States,
the United Kingdom, Europe and other countries.

ISBN: 978–0–230–11821–8

Library of Congress Cataloging-in-Publication Data

Adeola, Francis O., 1956–
    Hazardous wastes, industrial disasters, and environmental health
    risks : local and global environmental struggles / Francis O. Adeola.
        p. cm.
    Includes bibliographical references and index.
    ISBN 978–0–230–11821–8
        1. Hazardous wastes—Risk assessment. 2. Environmental health.
    3. Industrial accidents—Social aspects. I. Title.

TD1030.A326 2011
363.72'87—dc23   2011017519

A catalogue record of the book is available from the British Library.

Design by Newgen Imaging Systems (P) Ltd., Chennai, India.

First edition: October 2011

10 9 8 7 6 5 4 3 2 1

Transferred to Digital Printing in 2012

This book is dedicated to the loving memory of my beloved mother, Mrs. Mary Ọmọtinuwẹ Adeọla.

CONTENTS

List of Figures   ix

List of Tables   xi

Preface   xiii

Acknowledgments   xv

List of Abbreviations   xvii

**Part I   Hazardous Wastes, Disasters, and
             Health Risks**   1

1   Sociology of Hazardous Wastes, Disasters, and Risk   3

2   Hazardous and Toxic Wastes as a Social Problem   13

3   Taxonomy of Hazardous Wastes   25

**Part II   Electronic Waste, Persistent Organic
             Pollutants, and Health Hazards**   55

4   Electronic Waste: The Dark Side of the
    High-Tech Revolution   57

5   Environmental Health Risks of Persistent
    Organic Compounds   87

**Part III   Contaminated Communities and
             Regulatory Responses**   105

6   Communities Contaminated by Toxic Wastes and
    Industrial Disasters: Selected Cases   107

7   The Regulatory Frameworks   143

Part IV   Conclusion   **163**

8   Conclusion: Critical Environmental Justice Struggles   165

Appendix I: Principles of Environmental Justice (PEJ)   183

Notes   187

Bibliography   197

About the Author   227

Index   229

FIGURES

3.1 Hierarchy of Mercury Toxicity and a Wide Range of
Symptoms of Minamata Disease   41

3.2 Contact-Handled TRU Waste Being Processed
in an Airlock Room   49

3.3 A Shipment of TRU Waste Leaving Savannah
River Site (SRS) to Waste Isolation Pilot Plant (WIPP)   49

4.1 Mobile-Cellular Phone Penetration Rates   62

4.2 Distribution of Mobile-Cellular Subscriptions
in Africa, 2008   63

4.3a Unprotected E-Waste Dump Sites Posing Severe
Threats to Children in Less Developed Countries   65

4.3b E-Waste Contaminated Dump Site Posing a
Serious Danger to Children in Nigeria   66

4.4 A Model of Global Movement of E-Waste from
Generators to Receivers and Human Health Risks   67

4.5 A Nigerian Repairman Dismantling a CPU and
Monitor Attempting Repairs from Dismantled Parts
While Generating E-Waste from Unusable Parts   75

6.1 Poster of Lois and Missy Gibbs Promoting
Environmental Justice   111

6.2 Aerial Photograph of Abandoned Contaminated
Residential Structures Adjacent to Love Canal Site
Displaying Toxic Waste Rising to Ground Surface   112

6.3 Photograph of a Resident of Agriculture Street NPL
Site Displaying a Sign Showing the Extent of
Contamination and a Demand for Justice   116

6.4 Victims of Union Carbide Limited Toxic Gas
Explosion in Bhopal, India   121

6.5 Stockpile of Abandoned Toxic Chemicals inside
Derelict UCIL Warehouse in Bhopal, India   126

6.6 The Map of US-Mexico Border Regions
of *Maquiladoras*   136

Tables

3.1a Identity and Characterization of
      Hazardous/Toxic Waste   27
3.1b Department of Transportation Hazardous Materials
      Classification System   28
3.2   Sources and Types of Toxic Waste by
      Industry of Generators   29
3.3   Metal Ore and Metal Waste Production in the
      United States, 1960–1990   30
3.4   Top 20 Hazardous Substances and Substances
      Most Frequently Found in Completed
      Exposure Pathway (CEPs)   31
3.5   Toxic Synthetic Organic Compounds
      Commonly Present in Chemical Wastes and
      Their Health Effects   44
3.6   Major Sources of Nuclear Waste Types   51
4.1   Common Definitions of E-Waste   59
4.2   Global Growth in PC Sales and Use, Cell Phones, and
      Telephone Lines   61
4.3   Volume of Electronic Products Sold in the
      United States, 1975–2007   62
4.4   Hazardous Substances in PCs, Recyclability, and
      Their Potential Adverse Health Effects   73
5.1   Top-Priority POPs, Uses, and Their
      Adverse Health Effects   93
5.2   Top 10 Foods Most Contaminated with Specific POPs 96
5.3   POPs Pesticides Found in Total Diet Study in 1999 97
6.1   Eleven Towns and Populations Placed Under Health
      Monitoring within and around Contaminated Zones   128
7.1   Selected Federal Statutes Regulating Hazardous and
      Toxic Substances in the United States, 1965–2001   145
8.1   Major Milestones in the Evolution of
      Environmental Justice Movement   176

PREFACE

In the post–World War II period, modern societies have developed numerous heterogeneous synthetic organic compounds released into the environment and human habitats, either deliberately, inadvertently, or by industrial accidents. The threats posed by these contaminants and other hazardous wastes to human health and the health of other species in the environment are addressed in this book. Case studies of contaminated communities and the struggle for environmental justice are presented in a comparative cross-cultural perspective. These cases range from Love Canal, New York, and Agriculture Street, New Orleans, in the United States; the US-Mexico border; to Minamata, Japan; Seveso, Italy; Bhopal, India; Guiyu, China; and to Abidjan, Ivory Coast; and Koko, Nigeria; in West Africa.

This book focuses on hazardous and toxic waste releases, megaindustrial disasters, the consequent toxic contamination of communities, the environment, and the subsequent adverse health effects, injuries, deaths and destruction, and psychosocial problems. The book explains the emergence of the sociological study of risk, natural and technological disasters, and reviews the accumulated body of literature in the field till date. It integrates sociological perspectives with perspectives from other disciplines in the discussion of the problems posed by technological hazards both in the advanced industrialized societies and in the underdeveloped world. The book is unique in this respect.

The cross-cultural, comparative, and sociohistorical analyses of toxic waste dumping, industrial catastrophes, and morbidity and mortality effects on the exposed population also make the book quite unique. This book extends the body of knowledge through the innovative presentation of topics that have not been adequately covered in the extant sociological textbooks. The book starts with an introduction presenting the sociology of hazardous waste, risk, and disasters as a relatively new development with a growing passion and increasing volume of empirical research among social scientists.

Next, it describes how hazardous and toxic waste disposal, exposure, remediation, and proximate adverse health consequences have risen to the level of an endemic social problem both in the United States and around the world. Subsequently, the book delves into different classifications of hazardous wastes, the characteristics of each type of waste, and identifies what makes them especially dangerous to people and the environment. Other major topics addressed in the rest of the book are: electronic waste (e-waste) as a new species of trouble, the environmental and health risks of persistent organic pollutants (POPs), the case studies of contaminated communities within the United States and across the globe, the international flows of toxic e-waste, legislative imperatives and various attempts to regulate hazardous wastes, toxic releases, and industrial accidents in the United States, the European Union, and international agreements on transboundary movements of wastes. The issues of environmental justice, critical environmental justice struggles against persistent or enduring environmental inequalities, and key milestones in the evolution of environmental justice movement locally and internationally are addressed. The book is well illustrated with tables, charts, figures, and pictures to further stimulate the attention and interest of readers. It is written with both undergraduate and graduate students in mind. The book is also suitable for professionals, scholars, activists, and practitioners interested in gaining a better understanding of the intersection of toxic waste releases, human exposure to contaminants, and various health and psychosocial problems. This book is suitable as the main text or supplement for courses in environmental sociology, disaster studies, environmental studies, environmental policy, environmental health studies, social problems, waste management, global environmental studies, environmental regulations, ecology, and in social sciences and natural sciences.

ACKNOWLEDGMENTS

This book is a product of several years of rigorous research on the subjects of technological disasters, risks, environmental health, toxic wastes, and environmental injustice in the United States and around the globe. However, the whole idea of a book on the subject came during my sabbatical leave in the fall semester of 2009. For this, I thank the University of New Orleans and the entire Louisiana State University board of supervisors for approving my sabbatical leave for this endeavor. First and foremost, I must thank God Almighty for giving me the wisdom and strength to conceive an idea of writing and completing this book. Next, I must extend my gratitude to my family. During the entire period of developing and working on this book project, I received strong support from my wife and children. I wish to extend special thanks to my wife, Grace; my daughter Joyce Fadeke, who actually participated in drafting some of the figures; my young daughters Catherine and Bridget, who checked on me regularly to make sure I don't miss any mealtime; and my two sons Victor and Charles, who assumed the responsibility of yard work. I also wish to thank my good friend Jean Belkhir, editor of *Race, Gender & Class*, for his moral support, compassion, and keen interest for this book.

While there are numerous individuals, organizations, agencies, and groups that deserve recognition and praise, I am able to mention only a few names. My special gratitude and enormous thanks go to Christiane Charlemaine for facilitating the completion of this book. She devoted a considerable amount of time to formatting the transcript. Also, I would like to thank John Edens and William (Bill) Offhaus of the Special Collections Unit, University Archives, University at Buffalo; Ms. Lois M. Gibbs of the Center for Health, Environment and Justice; and Ms. Yuka Takamiya of the Basel Action Network for granting permission to use photos from their archives. In addition, I would like to thank the acquisition editor,

Burke Gerstenschlager, and editorial assistant, Kaylan Connally, for their suggestions, enthusiasm, and for working diligently with me to bring this book to production. I must also express my gratitude to my colleagues, who anonymously reviewed the earlier draft of the book manuscript and offered encouraging and constructive comments.

ABBREVIATIONS

ABSC	Acrylonitrile Butadiene Stryrene Copolymer
AHERA	Asbestos Hazard Emergency Response Act
ASHARA	Asbestos School Hazard Abatement Reauthorization Act
ATSDR	Agency for Toxic Substances and Disease Registry
BAN	Basel Action Network
BFR	Brominated Flame Retardants
CCASL	Concerned Citizens of Agriculture Street Landfill
CDC	Centers for Disease Control
CERCLA	Comprehensive Environmental Response Compensation and Liability Act
COP	Conference of Plenipotentiaries
CRJ	Commission for Racial Justice
CPUs	Central Processing Units
CTRs	Cathode Ray Tubes
DDT	Dichloro-diphenyl trichloroethane
EC	European Community
EJM	Environmental Justice Movement
EPA	Environmental Protection Agency
EPCRA	Emergency Planning and Community Right-to-Know Act
EOL	End of Life
EU	European Union
E-Waste	Electronic Waste
FAO	Food and Agriculture Organization
FDA	Food and Drug Administration
GAO	Government Accountability Office
GATT	General Agreement on Trade and Tariffs
GMOs	Genetically Modified Organisms
HLRW	High-Level Radioactive Waste
HCB	Hexachlorobenzene
HSWA	Hazardous and Solid Waste Amendment
IAEA	International Atomic Energy Agency
ICJB	International Campaign for Justice in Bhopal

ICMESA	Industrie Chimiche Medionali Societa Azionaria
INEERI	Indian National Environmental Engineering Research Institute
ITU	International Telecommunication Union
LDCs	Less Developed Countries
LDPFA	Land Disposal Program Flexibility Act
LERA	Lead Exposure Reduction Act
LLRW	Low-Level Radioactive Waste
LULUs	Locally Undesirable Land Uses
LUSTs	Leaking Underground Storage Tanks
MIC	Methyl Isocyamate
MDCs	More Developed Countries
MNCs	Multinational Corporations
MOSOP	Movement for the Survival of the Ogoni People
NACEC	North American Commission for Environmental Cooperation
NAFTA	North American Free Trade Agreement
NIMBY	Not in My Backyard
NGOs	Nongovernmental Organizations
NIEHS	National Institute of Environmental Health Sciences
NPL	National Priority List
NWC	Nuclear Weapons Complex
OAU	Organization of African Unity
OEMs	Original Equipment Manufacturers
OSHA	Occupational Safety and Health Act
PACs	Polycyclic Aromatic Compounds
PAHs	Polynuclear Aromatic Hydrocarbons
PBTCs	Persistent Bioaccumulative Toxic Chemicals
PCs	Personal Computers
PCBs	Polychlorinated Biphenyls
PEJ	Principles of Environmental Justice
PIC	Prior Informed Consent
POPs	Persistent Organic Pollutants
PPA	Pollution Prevention Act
PVC	Polyvinyl Chloride
RCRA	Resource Conservation and Recovery Act
SARA	Superfund Amendment and Reauthorization Act
SVTC	Silicon Valley Toxic Coalition
SWDA	Solid Waste Disposal Act
TCA	Trichloroethane
TCDD	Tetrachloro-Dibenzo-p-Dioxin

TCE	Trichloroethylene
TCP	Trichlorophenol
TRI	Toxics Release Inventory
TRUW	Trans-Uranic Waste
TSCA	Toxic Substances Control Act
UCC	Union Carbide Corporation
UCIL	Union Carbide India Limited
UNECE	United Nations Economic Commission for Europe
UNEP	United Nation Environmental Program
WEEE	Waste Electrical and Electronic Equipment
WIPP	Waste Isolation Processing Plant

PART I

HAZARDOUS WASTES, DISASTERS, AND HEALTH RISKS

Wastes are ubiquitous in society because every aspect of human activities involves the generation of different varieties of wastes. Directly related to metabolic processes in society are wastes ranging from solid wastes from households to extremely toxic industrial effluents and hazardous wastes from the industrial production of arrays of specialized products such as consumer electronics, computers, automobiles, petrochemicals, synthetic chemical compounds, plastics, and radioactive wastes (Tammemagi, 1999: 3). In modern societies, hazardous wastes cannot be avoided. Similarly, disasters from either natural or anthropogenic sources are increasing in frequency and appear inescapable. It is just a matter of time and place. There is a growing recognition among social scientists that hazardous wastes and industrial disasters pose immediate and long-term threats to the health and well-being of vulnerable populations along risky landscapes.

Part I consists of three chapters. Chapter 1 is the introduction where the emergence of the sociology of hazardous wastes, disasters, and risk is emphasized. Basic concepts in the study of natural and technological disasters and risks are presented. The wealth of literature on risks, hazardous wastes, and disasters of various etiologies are explored. A brief sketch of each of the subsequent chapters in the book is laid out.

In chapter 2, the problems associated with toxic waste releases and the basic characteristics of hazardous and toxic wastes are presented. The chapter also explains the distinction between natural hazards, technological hazards, and a complex of natural-technological (nat-tech) hazards, and how toxic wastes have been defined as an endemic social problem in societies all over the world. Chapter 3 covers the classification or taxonomy of wastes—by types, compounds, industry, sources, and nuclear or radioactivity. The adverse health effects of wastes from inorganic and synthetic organic compounds are also discussed.

1

SOCIOLOGY OF HAZARDOUS WASTES, DISASTERS, AND RISK

BACKGROUND

Social scientists did not pay adequate attention to the toxic and hazardous wastes generated in the course of industrial production in the United States during much of the nineteenth and twentieth centuries. There was a general failure to closely examine the interaction among the social, technological, and natural processes that often produce hazards and major disasters. Although systematic studies of disasters in the United States can be traced back to the 1940s-1950s (e.g., Lemons, 1957; Fritz and Williams, 1957), it was not until the late 1960s and early 1970s that the issue of toxic waste began to command the attention of many sociologists and other social scientists. In 1984, James F. Short particularly challenged sociologists to get involved in the study of risks, risk analysis, and disasters.[1] A series of catastrophic events during the period, involving extensive contamination and disruption of communities, loss of lives, and destruction of properties directed scholarly interest into the study of disasters and risks. Much of the emergent sociological research on disasters focused on short-term social responses to disasters of natural or human etiologies (Kreps, 1985; Kreps and Drabeck, 1996; Quarantelli and Dynes, 1977).

From the 1940s through the 1960s, the quest for capitalist expansion and rising affluence fueled the culture in which production and disposal of hazardous and nonhazardous by-products of industrial activities were viewed as unavoidable aspects of the capitalist system. Nature or the environment was viewed as capable of absorbing industrial effluents and replenishing itself without any long-term harm. Of course, the prevailing ideology then was the dominant Western worldview (DWW), according to which humans are fundamentally distinct from other creatures on earth by virtue of their culture, people are masters of their own destiny, the biosphere is vast and capable of providing unlimited possibilities for humans, and the history of humanity is one of progress that will continue with the invention and application of technologies (see Catton and Dunlap, 1980). Consequently, there was no stringent regulatory control of hazardous waste either at the federal or at the state level during this period.

Before stringent regulatory measures were implemented, an estimated 5 billion metric tons of highly toxic chemicals had been improperly disposed of in the United States between 1950 and 1975 (Cunningham, Cunningham, and Saigo, 2007). Understandably, much of the hazardous and toxic wastes grew in an exponential fashion in the United States during the period. Man-made disasters involving severe contamination of communities by previously unregulated toxic waste sites and industrial disasters increased during the 1970s and the 1980s with extensive media coverage, which sensitized the public to the health risks posed by exposure of people to xenobiotics. Among several cases of toxic contamination of communities are: the Love Canal (New York), Woburn (Massachusetts), Times Beach (Missouri), and Three Mile Island (Pennsylvania), in the United States; Seveso, Italy; Bhopal, India; and Chernobyl, Ukraine, chemical and nuclear disasters among other technological catastrophes around the world.[2] These cases clearly suggest that technological hazards are increasingly becoming a major source of risk to communities, human population, the environment, wildlife, and other living organisms.

Despite the impeccable record of economic prosperity and improved quality of life made possible by technological advancement since the industrial revolution, the dark side of technology is increasingly being revealed in the form of uncontrollable risks full of dreadful, uncertain, and unknown consequences. In recent years, the increasing frequency and severity of disasters of various etiologies around the globe have suggested the need for social scientists to gain a better understanding of these phenomena within the context of sustainability of society.

A disaster involves a process or an event with a combination of potentially destructive agents of natural, technological, or a complex of natural-technological (or nat-tech) etiology, affecting a population at risk in socially and physically produced conditions of vulnerability. Disasters are produced by interactions among three overlapping systems—the sociocultural systems, the built environment systems, and the biophysical environment in which society is embedded (Oliver-Smith, 1996; Youngman, 2009). Disaster generally affects different levels of society from individuals to groups, social organizations, institutions, and communities. A major disaster produces significant loss of property and lives, disruption of social institutions, and destruction of community infrastructures to the extent that vital functions of the society are disrupted or wiped out, which often results in physical and emotional distress and social disorganization of various sorts (i.e., the failure of social systems to meet their requisite needs).

Clearly, disasters have many consequences, including loss of lives and properties, overtaxing of available resources, disruption of daily activities, alteration of communities, and stresses and strains in the immediate postimpact phase, recovery phase, and in the long run. Victims of man-made disasters, such as toxic waste site contamination or industrial toxic exposures, have been found to be more likely to suffer from anxiety, depression, trauma, alienation, stigma, stress, and not being able to perform challenging tasks (Baum et al., 1992; Edelstein, 1988; Levi, Kocher, and Aboud, 2001).

Echoing the observation of several others, Girdner and Smith (2002: 2), in their book *Killing Me Softly*, contend that more than 40 years (i.e., 1969–2009) of environmental regulation in the United States has led to a failure to protect the environment from toxic contamination and its adverse health consequences. They further assert that the degradation of the environment should be considered a systemic problem—constituting a necessary and key component of the neoliberal capitalist political economy of industrial metabolism involving production, consumption, and waste. While their position is well taken, it should be mentioned, however, that environmental degradation is not limited to one political ideology or economic arrangement. Just as it is found in capitalist economies, hazardous and toxic waste proliferation and industrial accidents of megaproportions have also been documented in the socialist and communist societies such as the former Soviet Union, Russia, Eastern European countries, and Communist China (see Medvedev, 1990; Hamada and Osame, 1996). In the United States, the regulation of toxic substances and hazardous wastes has been an aspect of public policy generating tremendous conflict between the interests of polluting industries, the public, and government regulatory agencies. Curtailing industrial disasters has been problematic in recent decades, and in some cases, people have grown accustomed to high-risk technologies and what Perrow (1984) refers to as "normal accidents."

Although public attention has been called to the risk of exposure to toxic chemicals pervading the environment as early as 1962, with the publication of *Silent Spring* by Rachel Carson, a marine biologist, it was not until the 1970s that the public became seriously concerned about the risk associated with toxic waste contamination. A growing number of cases of communities contaminated by orphaned toxic waste sites, increased scientific knowledge about the adverse effects of toxins or xenobiotics on the environment and human bodies, and mass media coverage, have stimulated scholarly research and galvanized public attitudes concerning the issues of hazardous waste and public health. These have also energized grassroots antitoxic and environmental justice movements (EJMs; see Brown, 1992; Levin, 1982; Girdner and Smith, 2002; Bullard, 1990). In the United States, the leukemia cluster linked to the industrial toxins in the water supply of the residents of Woburn, Massachusetts; the landmark case of Love Canal, New York; and the Valley of the Drums and the cluster of cancers of various sites among the residents of the petrochemical industrial corridor of the Cancer Alley in Louisiana are cases of toxic contamination of communities across the country.[3]

There are several contested contaminated communities along the eighty-five-mile stretch along the lower Mississippi River corridor between Baton Rouge and New Orleans, known as the Cancer Corridor or Cancer Alley. This area has been a focal point for many environmental justice activists in the United States (see Bullard, 1990). More than 130 chemical plants and petroleum refineries with horrible environmental records are concentrated in the region. The chemical industry in Louisiana reports approximately 16,000 pounds of hazardous waste per capita in the state. Within the corridors are contaminated landscapes, homes, and human bodies—some of which have been diagnosed with environmentally induced illnesses, and several are yet to know their fate due to living and working within the vicinity of industrial plants within the lower Mississippi River corridor.[4]

Subsequent literature has reported that exposure to toxic and hazardous wastes from industrial activities and consumer goods in the environment has produced unintended adverse health effects—including cancers of various sites, birth defects, genetic defects, destruction of the immune system, respiratory problems, and reproductive disorders, among human populations as well as among wildlife (Epstein, Brown, and Pope, 1982; Lave and Upton, 1987; Thornton, 2000; Gould, Pellow, and Schnaiberg, 2008).

The issue of toxic waste exposure and associated health risks, particularly among working class and minority groups, became an extension of the civil rights movement, especially in the 1980s. The charges of environmental injustice and inequity in environmental regulation enforcement have been leveled against giant waste management companies, multinational petrochemical corporations, and environmental regulation authorities.[5] Grassroots mobilization among women and people of color, their shoe-leather popular epidemiology strategy, and public protests and demonstrations have been used effectively in advancing the goals of the EJM. Media coverage of a series of explosive toxic waste contamination of communities in the 1970s through the 1980s drew significant scholarly interest in the study of mass protests, public opposition to toxic waste facilities, the definition and perception of risks among the lay population and scientists, and the issues of civil rights and environmental justice.

Sociologists and other social scientists have now accumulated a huge volume of literature on the risk or danger posed by solid waste, hazardous and toxic wastes, nat-tech disasters, and other technological hazards to human health and the environment (see Levin, 1982; Edelstein, 1988; Freudenberg, 1984; Drabek, 1986; Brown and Mikkelsen, 1990; Brown, 1992; 2007; Baum, Fleming, and Singer, 1983; Freudenburg, 1997; Picou and Gill, 1996). The study of environmental health concerns was one of the focal points of environmental sociology, which emerged as a relatively young subdiscipline in the 1970s. Environmental sociologists have been involved in the study of technological risks, risk assessment, and risk perception in contemporary society. Using sociological tools—including sociological principles, theories, and methods—a growing number of sociologists have focused their attention on the risks of anthropogenic disasters relative to natural catastrophes.

As Tierney (1999: 216) observes, interest in the study of risks, hazards, and disasters of various etiologies has grown in the field of sociology since the 1980s. However, the development of a coherent overarching theoretical framework from which to explain and understand hazards and disasters remains a work in progress. A number of European social theorists, such as Ulrich Beck (1992, 1995, 1996, 1999), Anthony Giddens (1990), and Niklas Luhmann (1993), have each contributed abstract theoretical perspective addressing risks and technological hazards in modern society.

The concept of "risk society" was introduced by Ulrich Beck to describe a phase of development of modern society in which the risks introduced by the momentum of technological and social innovation increasingly elude the management and control of the protective institutions of industrial society. According to Beck (1996: 27), the transformation from industrial society to "risk society" involves two distinct phases. In the first phase, self-endangerment and adverse consequences are systematically produced but are not the subject of public debate, or at the core of any political rift. This was the case in the United States for the most part of the 1940s through the 1950s.

In the second phase, a totally different situation develops especially as the hazards and risks of industrial society now dominate public, political, and private debates. The institutions now acknowledge and legitimize hazards that are not controllable in this epoch of reflexivity. Beck (1996: 31) suggests:

> The entry into risk society occurs at the moment when the hazards which are now decided and produced by society undermine and/or cancel the established safety systems of the provident state's existing risk calculations. In contrast to early industrial risks, nuclear, chemical, ecological, and genetic engineering risks are: (1) unlimited in terms of time and space, (2) not accountable according to the established rules of causality, blame, and liability, (3) impossible to be fully compensated or insured against.... For instance, the victims of Chernobyl and *Bhopal* are today, years after the catastrophes, not even all born yet.

Gradually, an industrial society has involuntarily mutated into a risk society through its own systematically produced hazards that are now being shifted or spread to the rest of the world through globalization. Tierney (1999) points out that the texts of Ulrich Beck and his European colleagues are highly abstract and generally without empirical grounding. Furthermore, their analyses are focused on risks associated with technologies in contemporary Western industrialized societies without much attention being paid to risks and disasters encountered in the non-Western world.

As mentioned earlier, scholarly research endeavors on risks and disasters have grown phenomenally since the 1970s. With regard to disasters brought about by orphaned waste sites, the works of Adeline Levin, Michael Edelstein, and Phil Brown are very significant. Levin's (1982) book, *Love Canal: Science, Politics, and People* eloquently chronicled the story of buried drums of toxic chemicals that contaminated the homes, schools, and bodies of the residents of a middle-class neighborhood of Niagara Falls, New York. The consequent environmental disaster, popular epidemiology, community mobilization, the organization of the antitoxic waste movement, and the eventual resolution are well documented.[5] In *Contaminated Communities: The Social and Psychological Impacts of Residential Toxic Exposure*, Edelstein (1988) observed a water-contamination event in Legler, New Jersey, detailing the psychosocial impacts, including the feelings of stress and helplessness, along with the physical effects of exposure to toxic contamination. Phil Brown and E. J. Mikkelson meticulously narrated and analyzed the plights of the residents of Woburn, Massachusetts, in their book *No Safe Place: Toxic Waste,*

Leukemia and Community Action. The idea of popular epidemiology was explained in this text and other subsequent publications (e.g., see Brown, 1992, 2007).

In a comparative cross-national approach, Michael Reich's (1991) *Toxic Politics: Responding to Chemical Disasters* presents three major disasters including the Michigan polybrominated biphenyl (PBB) contamination in cattle feed in 1973, the polychlorinated biphenyl (PCB) contamination of cooking oil directly consumed by humans in western Japan in 1968, and the dioxin explosion and contamination from a factory in Seveso, Italy, in 1976. He explores the ways in which the victims sought redress and shows that victims of toxic contamination seek redress through politics by making the problem public, organizing groups for collective action, and by mobilizing political allies. These findings are consistent across different cultures. He further shows that this process often creates its own conflicts and costs, its own process of victimization, and how the politics of toxic contamination or disaster can become as toxic and corrosive for the victims as the disruptive agents themselves (Reich, 1991). Similar observations have been made earlier by Kai Erikson (1976, 1994) in *Everything in Its Path: Destruction of Community in the Buffalo Creek Flood,* and in *A New Species of Trouble: Explorations in Disaster, Trauma, and Community.* Even though the Buffalo Creek disaster was not a toxic crisis, the work of Erikson served as one of the first book-length accounts of an anthropogenic ecological disaster in the United States by a sociologist.

More than 30 years ago, Eckardt C. Beck, a regional administrator for the Environmental Protection Agency (EPA), offered a chilling statement concerning Love Canal as being merely the tip of the iceberg: "We suspect that there are hundreds of such chemical dumpsites across this nation." Confirming his suspicion since the discovery of the Love Canal toxic contamination, thousands of hazardous waste sites ranging from local dumps to accidental spills have been found throughout the United States. The list of the most dangerous sites known as the National Priority List (NPL) compiled by the EPA has grown exponentially since the Love Canal saga. The EPA is charged with the responsibility of identifying, investigating, and cleaning up of such sites, and finding the responsible party for sanction.[6]

PURPOSE OF THIS BOOK

This book focuses on systematic and industrial disasters that release hazardous and toxic wastes into the biophysical and built environments culminating in contamination of communities and the subsequent adverse health consequences among the exposed populations. The deleterious effects of toxic contaminants on populations and the environment are of particular concern. Several aspects of technological disasters resulting from the interface of human sociocultural systems, technological complexes, and the biophysical environment are addressed in this book. Hazardous waste classification and identification of major sources of release of toxic contaminants to communities are addressed. The book also examines industrial or technological disasters and their impacts on the host communities within the United States and in other parts of the world. A number of case studies of communities contaminated due to exposure to xenobiotics via hazardous or toxic waste sites and through catastrophic accidental releases are presented. The comparative cross-cultural case studies presented make this book especially appealing to students, researchers, and practitioners interested in understanding the plights and struggles of people exposed to toxic contaminants across cultures. The question as to whether human beings exhibit similar reactions when faced with the reality of toxic contamination and actual or potential adverse health and well-being across cultures is addressed in this book.

ORGANIZATION OF THE BOOK

The book is organized into four major parts. Part I consists of three chapters (chapters 1–3), with chapter 1 devoted to the introduction and the emerging literature on the sociology of technological disasters and risks; an overview of what to expect in the remaining chapters of the book is also laid out in this chapter. Chapter 2 presents toxic waste and community contamination as a social problem, and chapter 3 is devoted to the classification and properties of hazardous waste. Attention is focused on different types of toxic wastes, their sources, and their adverse health effects on humans and the environment.

There are two chapters in part II (chapters 4 and 5). In chapter 4, the problem of electronic waste (e-waste) as the latest plague of our times, crossing national and cross-national geopolitical boundaries, and its associated environmental health concerns are discussed. The scope of the problem, including the huge volume of e-waste being generated annually, and the issue of ineffective control of transboundary movements of these materials are addressed. The flow of e-waste along the paths of least resistance from the global North to the global South is addressed within the framework of global environmental injustice. Furthermore, the health and environmental impacts of e-waste from cradle to grave are addressed in a comparative cross-national approach. Chapter 5 covers the characteristics, nature, and potency of persistent organic pollutants (POPs). Their adverse effects on human population and the environment are described with specific illustrations in this chapter.

In part III, there are two chapters (chapters 6 and 7). Chapter 6 focuses on specific cases of contaminated communities either by exposure to toxic waste dumps or by accidental releases of toxic chemical compounds. The cases of community toxic contamination at Bhopal, India; Seveso, Italy; and the Love Canal (New York), Woburn (Massachusetts), and Agriculture Street (New Orleans), respectively in the United States, are presented. Pertinent environmental legislation that forms the legal and regulatory frameworks addressing the problems of toxic wastes and industrial hazards in society is the main focus of chapter 7. Part IV is the last segment of the book and it consists of the concluding chapter (chapter 8) devoted to environmental justice struggles. In chapter 8, the origins, inception, evolution, and diffusion of environmental justice movements within the United States and globally are discussed with emphasis on milestones of the EJM. The book ends with an appendix and bibliography or references.

2

HAZARDOUS AND TOXIC WASTES
AS A SOCIAL PROBLEM

INTRODUCTION

The central focus of this chapter is on different categories of toxic wastes and their etiology in the environment. First, pertinent background literature on hazardous and toxic waste is reviewed. Second, the concepts of "hazardous waste" and "toxic waste" are defined. Subsequently, to facilitate our understanding of the nature of toxic waste vis-à vis other hazards, the distinction between toxic/technological hazards and natural disasters is clearly presented; how toxic wastes have become prominent as a major social problem that evolved through social construction is also discussed in the chapter. The polarizing nature of toxic waste and other technological disasters and their chronic effects at various levels of society are elucidated as well.

There has been a phenomenal acceleration in the production and distribution of toxic chemicals and hazardous wastes over the past two centuries in the advanced industrialized nations of the global North. The Industrial Revolution, the scientific and technological breakthroughs in the synthesis and production of heterogeneous chemical compounds, and the rise of the consumerist culture among the growing population are directly linked to the growth of hazardous wastes. For the most part, hazardous and toxic wastes are the by-products of a highly technological society, or to use Ulrich Beck's term, a "risk society." From modern agricultural production to industrial activities, from private households to public arena, and from commercial to military operations, hazardous wastes are ubiquitous in modern society. Toxic wastes and their deleterious effects on human health and the environment are now of growing concern. As noted by Lichtveld and Johnson (1993), concerns about toxic and hazardous wastes continue to increase globally. Generally, people are concerned about hazardous waste sites, accidental releases of toxic substances, expensive clean-up costs, property-value depreciation, stigma attached and other social and psychological costs, adverse ecological effects, and human-health diminution (Adeola, 2000a; Blackman Jr., 1996; Lave and Upton, 1987; Thornton, 2000; Brown, 2001).

Public consciousness, knowledge, and concerns about the dangers of toxic contamination have increased in the United States, Europe, and other parts of the world in the post–World War II era (Carson, 1962; Freudenberg, 1984; Erikson, 1994; Setterberg and Shavelson, 1993; Edelstein, 1988).Yet, policies and programs aimed at curbing toxic waste production, release, and dispersion remain grossly inadequate. The problem of toxic waste control is exacerbated by the lack of a universally acceptable system of waste classification. The grouping and characterization of wastes vary by region, country, and industry. Even government agencies differ in their schemes of classification of hazardous substances.

By their properties including corrosivity, ignitability, irritability, reactivity, and toxicity, as established under the US Resource Conservation and Recovery Act (RCRA), hazardous wastes pose significant danger to human and nonhuman species in the environment. In recognition of how widespread technologically induced risks and environmental hazards are, Ulrich Beck (1992) coined the term "risk society" to describe the contemporary advanced industrial societies where the social production of wealth is inextricably coupled with the social production of hazards and associated risks. Thus, "risk society" connotes an epoch in which the "dark sides" of modernity are increasingly dominant in social discourse; self-endangerment and the ruination of nature or massive environmental degradation have become prominent among the major concerns (Beck, 1995). Toxic releases and psychophysiological problems associated with modernity and its technological accoutrement have been referred to as "a new species of trouble" (Erikson, 1994). Acknowledging the ubiquity of toxins in the environment, Brown and Mikkelsen (1990) concluded that there is no such thing as a safe place. As noted by Epstein, Brown, and Pope (1982: 37), the materials and functions that are generally taken for granted as basic aspects of everyday life in modern society depend upon a geometric increase, proliferation, and distribution of an array of heterogeneous toxic chemicals.

The post–World War II era in particular has witnessed an unparalleled acceleration of the production and distribution of synthetic toxic organic chemicals—including pesticides (e.g., dichlorodiphenyltrichloroethane (DDT)), dioxin, polychlorinated biphenyls (PCBs), synthetic fibers, and other chlorine products—all with significant adverse impacts (Carson, 1962; Epstein et al., 1982; Brown, 2001; LaGrega et al., 2001; Field and Field, 2002). From the 1930s to the present, the production of synthetic organic chemicals in the United States has increased more than 30-fold. Currently, there are more than 70,000 industrial chemicals being sold in the market (Thornton, 2000; Cohen and O'Connor, 1990). Brown (2001: 131) contends that the exact number of manufactured chemicals presently in society is unknown, and especially with the introduction of synthetic organic compounds, the number of chemicals in use is estimated at 100,000. Over 1,000 toxic chemicals in various discarded electronic gadgets are dispersed all over the world in the form of electronic waste (e-waste).[1] According to Ulrich Beck, these hazards associated with industrialization...do not assail us like a faith; rather they are a product of human hands and minds, of the link between technical knowledge and the economic utility calculus. Unlike wars, these risks enter the world peacefully and they thrive in the centers of rationality, science, and wealth.[2]

Today's toxic pollution problem consists of many interrelated aspects that can be summed up as the "toxic cycle" or the "treadmill of toxics." Cohen and O'Connor define a "toxic cycle" as the production, use, and disposal of toxic chemical products considered essential in modern society. The "treadmill of toxics" on the other hand refers to the societal inclination toward the ever-increasing demand and supply of toxic materials as a major aspect of industrial capitalism (see Schnaiberg and Gould, 1994). The petrochemical industry has created both the toxic cycle and treadmill of toxics by shaping public demand through the production and advertisement of large volumes of synthetic chemicals offered as substitutes for naturally occurring products. Consistent with Schnaiberg and Gould's (1994) treadmill of production, the treadmill of toxics is rooted in a network of regional, national, and transnational corporations, who place profits above human health and environmental protection. Increased toxic waste generation is a major aspect of the planned obsolescent scheme of the treadmill of production and consumption now pervasive in advanced industrial societies.

The Pollution Prevention Act (PPA) of 1990 mandates facilities to report information in a toxics release inventory (TRI) about quantities of toxic chemicals they manage in waste, both on-site and off-site, including amounts reported as recycled, burned for energy recovery, treated, disposed, or otherwise released. Recent TRI data from the US Environmental Protection Agency (EPA) show that industries in the United States produced and released 4.1 billion pounds of toxic waste in 2007 (the latest year for which data are available), which represents a significant drop from the 7.8 billion pounds reported in 1999 (EPA, 2009; 2001: E-2–3). Metal mining industry, which started reporting in 1998, accounted for the largest total releases, followed by electric utilities, chemical manufacturing, and the paper industry sectors, respectively. This is consistent with the 1999 TRI data showing percentage distribution of releases by industry sectors: metal mining dominated by 51.2 percent; the original chemical manufacturing industries accounted for 29.9 percent; electric generating facilities accounted for 15 percent; RCRA subtitle C, treatment, storage, or disposal (TSD) and solvent recovery facilities, accounted for 3.7 percent; and others including coal mining, chemical wholesale distributors, and petroleum terminals and bulk storage facilities accounted for the remaining small percentage of total toxic releases, respectively. Among the limitations of TRI is the fact that small generators of toxic waste are not required to report. The self-reporting required of large generators may be questioned for data reliability and validity because companies may underreport their actual volume of toxic releases. As noted by EPA (2009: 2), TRI does not include non-production-related releases, which encompass releases due to natural disasters (as during the flood due to Hurricane Katrina), accidental leaks, or other one-time occurrences that are not considered part of the routine process.

The United States is the world's largest generator of hazardous wastes. According to Epstein et al. (1982), the majority(if not all) of Americans carry substantial burdens of lead, pesticides, polychlorinated biphenyls (PCBs), and other harmful chemicals in their bodies as a result of exposure to polluted air, soil, and water, food chain contaminants, and due to direct use of toxic consumer products. In some cases, the source of exposure to toxic elements are the homes and work environment where thousands of people across America are directly exposed to lead, asbestos, radon, and other xenobiotics (Edelstein and Makofske, 1998). Thus, toxic chemicals are present in the air we breathe, the land we tread upon, and the water we drink. Toxic wastes are present in our workplaces, schools, on our farms and gardens, on lawns in our neighborhoods, and in our bodies (Brown, 2001; O'Connor, 1990; Thornton, 2000).

DEFINING HAZARDOUS AND TOXIC WASTES

What is hazardous or toxic waste? It is apropos to gain an understanding of what constitutes waste and what are the different types of toxic waste. Generally, waste is defined as any unwanted materials either in solid, liquid, or contained gaseous form discarded by being disposed off, buried, burned or incinerated, or recycled (EPA, 1996). Although the terms "hazardous" and "toxic" are frequently used interchangeably in the literature, these terms, however, are not quite synonymous. In simple terms, "hazardous" connotes harmful or dangerous materials, while "toxic" implies lethal or poisonous substances (Cunningham and Saigo, 1999). Hazardous waste represents a broad category of wastes in solid, liquid, or containerized gaseous form known to be harmful to humans and other species due to their ignitable, corrosive, explosive, toxic, carcinogenic, mutagenic, or teratogenic characteristics (Miller, 2001; EPA, 1996; Epstein et al., 1982; Nebel and Wright, 2000). Nebel and Wright (2000) define hazardous material as anything that can cause injury, disease, or death; damage to property; or significant destruction of the biophysical environment. The US RCRA defines hazardous waste as any solid waste or a combination of solid wastes that, due to their quantity, physical, chemical, or infectious properties, may cause or significantly promote an increase in the number of deaths or an increase in serious irreversible, or incapacitating reversible, illness; or pose a serious present or potential danger to human health or the environment when improperly treated, stored, transported, or disposed of, or otherwise managed (Tammemagi, 1999: 66–67). Toxic waste represents a subcategory of hazardous waste that may cause death or permanent (irreversible) or severe (reversible) impairment to human and nonhuman species upon contact (LaGrega et al., 2001; Tammemagi, 1999; Asante-Duah, 1993).

Another attribute of toxic wastes involves their degree of toxicity. Acute toxicity and chronic toxicity may occur upon exposure of humans and other living organisms to toxic wastes. The former results in either sudden death or illness shortly after a single exposure to a toxic substance (such as in the case of Bhopal, India, and Seveso, Italy), while the latter may take a long latency period to manifest, as was the case in Woburn, Massachusetts, Love Canal, New York, and numerous other contaminated communities across the United States (see Levine, 1982; Brown and Mikkelsen, 1990; Edelstein, 1988). Most of this type of waste now represents the dark side of modern science and technology—the key elements of a risk society. As noted by Field and Field (2002: 333), hazardous and toxic substances possess traits that pose unique difficulties for monitoring and control, both at the national and the cross-national levels, including the following:

1. The fact that hundreds of new chemicals are developed and introduced into the market annually, and that more than 300 million

tons of chemicals are produced, makes tracking the substances being used and the quantity of substances produced extremely difficult.

2. Given the fact that thousands of chemicals are in use, each with different chemical and physical properties, it becomes virtually impossible to be completely informed about their toxicity and their adverse effects on humans and other species in the ecosystem.

3. Monitoring and tracking small-quantity generators of heterogeneous toxic chemicals is especially difficult. Most small-quantity generators are extremely difficult to police due to their surreptitious disposal practices, often known as "mid-night dumping."

4. Health problems and other damages due to exposure to hazardous/toxic materials generally take many years to manifest; thus, the time lag and the chronic nature of the damages often blur the direct "cause" and "effect" connection.

How different are natural hazards from technological hazards? This is the important question addressed next in the following section.

NATURAL HAZARDS VS. TECHNOLOGICAL HAZARDS

Hazards of various kinds have existed in human society since the dawn of civilization and perhaps since time immemorial. People have struggled to cope with and adapt to numerous hazards of both natural and anthropogenic etiologies since the first human settlements and invention of technology. Natural hazards are somewhat predictable, accepted as "an act of God"; technological hazards are mostly unpredictable, but with an identifiable culpable party. Furthermore, natural hazards are considered to be amenable to risk analysis, which estimates the probability of occurrence of such events and their likely outcomes. In addition, natural hazard processes are correlated in several ways—for example, hurricanes, flooding, and coastal erosion are inextricably linked.

Social scientists have distinguished technological disasters from natural calamities. Unlike natural disasters, toxic waste contamination and the consequent health and socio-psychological sequelae are anthropogenic in origin (Picou, 2000; Erikson, 1994; Edelstein, 1988; Baum, 1987). A number of studies have shown that human-induced toxic disasters are technological in origin and often yield greater chronic (long-term) physical, biological, and psycho-social problems of epic proportions (see Edelstein, 1988: 6; Baum, 1987;

Baum and Fleming, 1993; Picou, 2000). Technological disasters such as toxic releases, oil spills, and any other suddenly imposed technological disruptions tend to create a social condition of uncertainty, anger, anxiety, frustration, blame, depression, isolation, a loss of control, and a distrust of governmental authority and other related agencies (Edelstein, 1988; Tucker, 1995; Picou, 2000; Adeola, 2001; Levine, 1982; Hallman and Wandersman, 1992). Chronic environmental disasters are more difficult for people to cope with relative to acute environmental problems due to systemic natural forces. As noted by Erikson (1994: 144, 151):

> Toxic poisons provoke a special dread because they contaminate rather than merely damage; they are stealthy (silent killers) and deceive the body's alarm systems; they pollute, taint, and befoul rather than just create a wreckage; and because they can be absorbed into the very tissues of the body and crouch there for years, even generations, before doing their deadly work.

In an expert panel workshop on the psychological responses to hazardous substances, convened by the Agency for Toxic Substances and Disease Registry (ATSDR) in 1995, the unique psychosocial aspects of exposure to man-made toxic substances were summarized thus: "Unlike the damage and injuries inflicted by natural disasters, many toxic substances are invisible to the senses; this invisibility results in feelings of uncertainty. People cannot be sure without instrumentation if they have been exposed to a toxin and how much they have been exposed. Furthermore, due to significant time lag between exposure and manifestation of a chronic illness related to exposure, it is extremely difficult to establish a direct cause and effect relationship between past exposure and subsequent illness" (Tucker, 1995: 2).

Social conflict is often endemic in cases of toxic disasters while cases of natural disasters are less contentious or litigious (Adeola, 2000a). The uncertainty, the divergent meanings, and the claims and counterclaims that characterize technological disasters such as toxic waste contamination often yield a "corrosive community" in contrast to an altruistic, compassionate, and "therapeutic community" that typically emerges spontaneously as people mobilize social capital in response to natural disasters (see Drabeck, 1986; Freudenburg, 1997; Cuthberson and Nigg, 1987; Gill, 2007). The chronic toxic effects often entail significant biological dysfunctions including the destruction of vital body organs, the central nervous system, the gastrointestinal tract, and genetic material. Different types of cancers, birth defects, embryo toxicity, mutations, and other dreadful diseases are associated with long-term exposure to toxic elements in the environment. Unfortunately, tracing the specific environmental etiology of these health problems over time is often difficult. For natural disasters, however, the impacts are typically of short duration, identifiable, and recovery generally occurs within a reasonable time span.

There are some events involving a juxtaposition of the characteristics of both natural disaster and human system failure. For instance, the breaches of several levees and other flood protection structures, which led to a massive flooding of 80 percent of the city of New Orleans by Hurricane Katrina in August of 2005, represent a complex of a natural-technological (nat-tech) disaster. Even though the storm was an act of nature, the subsequent death, suffering, and destruction are largely attributed to the failures of the human agency at different levels, making the flood due to Hurricane Katrina an un-natural catastrophe. Several toxic waste sites placed on the National Priority List (NPL) by the US EPA in Orleans Parish formed the basic ingredients of the toxic gumbo released by the floodwaters into thousands of homes and offices.

Toxic Waste as an Endemic Social Problem

A social problem implies any existing undesirable condition adversely affecting a substantial segment of the society, viewed apprehensively and distastefully by an influential group who believes this condition can be mitigated or eliminated through concerted collective efforts (Zastrow, 2000). The social problem generally evolves through the process of social construction involving claims making, counterclaims, and a resolution (see Hanningan, 1995; Greider and Garkovich, 1994; Albrecht and Amey, 1999). The major factors essential for the successful construction of an environmental issue as a social problem are: (1) an influential group or claims promoter organizing and making a claim or pursuing a cause concerning a putative condition; (2) a community consensus regarding the existence and nature of the problem upon which a claim is centered; (3) a scientific endorsement of the claim; (4) significant media attention and coverage; (5) a dramatization of the problem in symbolic and visual terms; (6) economic and sociopolitical incentives; and (7) an institutional sponsor that can ensure legitimacy and continuity of the claim(s) (Hannigan, 1995; Faupel, Bailey, and Griffin, 1991). These factors have played significant roles in the process of naming and framing toxic waste as a major social problem.

The issues concerning toxic wastes have been galvanized by extensive media coverage of toxic waste-related disasters in recent years (Gill and Picou, 1998; Faupel et al., 1991). A number of toxic chemical disasters that have made national and international headlines are: the Union Carbide chemical releases in Bhopal, India; the nuclear meltdown at Chernobyl, Soviet Union (now Ukraine); the massive oil spill by Exxon Valdez in Prince William Sounds, Alaska, and the severe and chronic toxic chemical contamination at the Love Canal, New York, in the United States; mercury poisoning in Minamata Bay, Japan; dioxin contamination in Seveso, Italy; the exposure of thousands of migrant farmworkers to pesticides, in the United States; toxic chemical contamination in Koko, Nigeria; e-waste contamination in Guiyu, China; and several others.

The proliferation of antitoxic movements, the not-in-my-backyard (NIMBY) syndrome, and powerful environmental organizations have been instrumental in framing toxic or hazardous waste issues as a social problem at the local grassroots, national, and international levels. The new social movement for environmental justice has played vital roles in raising people's consciousness and activism concerning toxic waste problems at the local, national, and global levels. Mass media exposure coupled with increased knowledge about acute and chronic health effects of toxic wastes in the environment have contributed to the conception of toxic waste as a major social problem in today's society. This is particularly so because the most toxic waste contamination episodes are attributed to humans' reckless activities rather than natural forces as mentioned earlier (Gill and Picou, 1998; Edelstein, 1988; Baum, Fleming, and Singer, 1983). Most toxic wastes are externalities of industrial production in which the costs of production are shifted to innocent bystanders—mostly powerless and disenfranchised groups in society (see Adeola, 2001, 2000b).

Several cases of community contamination such as the Love Canal, New York; Times Beach, Missouri; Agriculture Street, New Orleans; Louisiana; a number of communities along the cancer corridors of Louisiana and Texas; Woburn, Massachusetts; Three Mile Island, Pennsylvania; Chernobyl, Ukraine; Koko, Nigeria; Exxon Valdez massive oil spills in Prince William Sound, Alaska; and many more across the world have served as eye-openers concerning the deleterious consequences of technological disasters and toxic wastes on humans and the environment (see Adeola, 2000a; Edelstein, 1988; Picou, 2000; Brown and Mikkelsen, 1989; Brown, 2007). These catastrophes have convinced the general public about the significant dangers associated with the production, distribution, use, and improper disposal of toxic substances. Most of these cases elicit powerful images of toxic chemicals causing severe ecological destruction, community disruptions or disorganization, death, or acute and chronic toxicity affecting human and nonhuman species (Edelstein, 1988; Levine, 1982; Hernan, 2010). Most importantly, these cases have exposed the frailty of modern science and technology, leading to a substantial erosion of public trust in the system and in the government most especially.

There is a growing consensus among experts, scientists, government, industry, and the lay public that toxic waste releases in the environment constitute a significant social problem that is endemic to many communities throughout the global North and the underdeveloped societies of the global South. Available data indicate that the spread of toxic chemicals in the environment is causing a public health crisis of epic proportions; for example, different types of cancers, respiratory problems, chemical hypersensitivity syndrome, sterility, birth defects, embryo toxicity, nervous system dysfunctions, vital organ dysfunctions, mutations, and retardation are among the litany of health problems associated with toxic waste pollution (Adeola, 1994; Arcury and Quandt, 1998; Brown, 2001; Thomas et al., 1998; Tsoukala, 1998; Rosenthal, 2004). Annually, Americans spend millions of dollars to address these adverse health conditions. With our understanding that toxic wastes and associated health issues constitute a serious social problem affecting a large segment of the society, it is crucial to gain familiarity with toxic waste classification. The next chapter is devoted to the issue of classification of toxic waste with emphasis on the United States.

CHAPTER SUMMARY

In this chapter, several aspects of toxic wastes have been considered, including the definition and social construction of toxic wastes as a social problem. The nature of risks associated with toxic wastes and other technological hazards calls for our immediate concerns. While natural disasters involve a loss of control over processes perceived to be uncontrollable in the first place, toxic disasters involve a loss of control over conditions or processes perceived to be controllable (Baum et al., 1983). Thus, parties to blame are often identified in cases of toxic contamination and other technological accidents, which sometimes creates conflict at different levels and a corrosive community. With increasing knowledge and public awareness of the problem of toxic waste releases, scientists, environmental activists, lay persons, and the media are among the key players in the social construction of toxic waste as a serious problem.

The problems of toxic waste releases and management are serious in advanced industrial societies of the global North, who are constantly seeking cheaper routes of waste disposal. Hazardous waste dumping is especially endemic in many Third World societies of the global South with lax regulation and less capacity or technical knowhow for toxic waste management. Even though the Emergency Planning and Community Right-to-Know Act (EPCRA) requires large generators of toxic waste to declare all their toxic releases annually in the United States, underreporting by these large generators and significant amounts of toxic releases from small generators and other sources not covered by the legislation remain as serious threats to the health and well-being of the population and the environment. The topic of waste classification and properties or characteristics of different types of waste are addressed in chapter 3 that follows.

3

Taxonomy of Hazardous Wastes

This chapter is devoted to the classification of waste. Different classes of waste and their intrinsic properties are examined. Among hazardous wastes, broad categories of inorganic compounds or heavy metals, specific organic compounds, and biomedical and radioactive wastes are presented along with their adverse effects on health and the environment.

Classification of Waste

In order to develop effective strategies to manage and control wastes and at the same time protect human health and the environment, it is imperative to identify and classify toxic and hazardous wastes properly. There have been numerous attempts at the classification of hazardous wastes. Toxic wastes are usually treated as a subcategory of hazardous wastes (Asante-Duah, 1993; Epstein et al., 1982; LaGrega et al., 2001). One crude approach is to group wastes according to the degree of risk they pose to humans and the environment. Thus, wastes are put into high-, intermediate-, and low-risk categories. High-risk wastes have the properties of being highly toxic, persistent, mobile, ignitable, and bioaccumulative. Examples of this type of wastes include chlorinated solvents, persistent organic pollutants (POPs), heavy metals such as lead and cyanide wastes, and polychlorinated biphenyl (PCB) wastes. Intermediate-risk wastes are mostly insoluble and have low mobility, for example, metal hydroxide sludges. Low-risk wastes generally include high-volume, nontoxic, and malodorous wastes, for example, municipal solid waste (see Asante-Duah, 1993). A waste is considered hazardous or toxic if it meets any of the four properties set by the Environmental Protection Agency (EPA) under the provisions of the Resource Conservation and Recovery Act (RCRA), that is, ignitability, corrosivity, reactivity, and toxicity.

Any liquid with a flash point of less than 140 degrees Fahrenheit or a solid that is capable of causing fire either through friction or through absorption of atmospheric moisture, or can undergo spontaneous chemical change resulting in fire, is defined as an ignitable or flammable waste. Ignitable wastes are given an EPA hazardous waste number of D001. Corrosivity is determined by using a pH scale. Thus, any waste with a pH less than or equal to 2 or a pH equal to or greater than 12.5 is considered a corrosive waste. The EPA number D002 classifies corrosive wastes. Reactive wastes are materials that are unstable and can undergo violent chemical change without detonating, can react violently with water to form possible explosive mixtures, or may generate poisonous gases. They are coded as D003 waste by the EPA. Acute toxicity is defined as the lethal dose of a chemical that takes less than 50 mg/kg of body weight to kill 50 percent of the population exposed (< LD50), for example, hydrogen cyanide or hydrogen sulfide. Other toxicity thresholds have been established by the EPA based on oral, dermal, and inhalation toxicity (see Tammemagi, 1999: 80). Thus, any waste that may cause death or severe injuries to the exposed victims is considered a toxic waste (see table 3.1a).

Hazardous wastes are produced and transported across the nation and in some cases cross-nationally. There are numerous cases of shipment and dumping of toxic wastes that originated from affluent nations of the global North to impoverished or underdeveloped societies of the global South. Within the country, the US Department of Transportation (DOT) has developed a system of classification of hazardous and toxic substances transported within the interstate commerce. Table 3.1b presents the DOT classification scheme and representative hazardous substances the agency regulates. Flammable and combustible substances are involved in most cases of accidents. Hazardous materials that are explosive, poisonous, corrosive, and radioactive are of greatest concern.

Toxic wastes have also been classified by source, industry, chemical composition and degree of toxicity, persistence in the environment, radioactivity, and public health threats (see Epstein et al., 1982; Tammemagi, 1999; Asante-Duah, 1993; McGinn, 2000; LaGrega et al., 2001). For the present purpose, toxic wastes are classified by type, that is, inorganic and organic compounds, industry, and sources. The EPA has developed a system of classification of hazardous wastes by industry. The typical industry and the types of toxic and hazardous wastes generated are shown in table 3.2. The most hazardous and toxic wastes are produced by chemical manufacturing, petroleum,

Table 3.1a   Identity and Characterization of Hazardous/Toxic Waste

Identity/Properties	Quality/Characteristics
Corrosivity	Aqueous with pH of 2 or less; liquid with pH of 12.5 or higher that corrodes steel.
Explosivity	Any waste that forms a potential explosive mixture, reacts, or detonates at 25°C and 1 atmosphere.
Flammability/ Ignitability	Any liquid waste that inflammes or ignites at 60°C or nonliquid waste that inflammates under 25°C and 1 atmosphere; produces fire by friction, humidity absorption, or by spontaneous chemical changes; and waste that acts as an oxidizer. The oxygen released promotes the combustion intensity of other materials.
Mutagenicity	Any waste capable of causing aberration in DNA structure or mutation of genetic material.
Pathogenicity	Any waste that contains a harmful microorganism or a toxic biological agent that may cause harm, illness, or death.
Reactivity	Wastes that are unstable that react violently without detonation; react violently when mixed with water; and generate gases, vapor, and toxic smoke containing cyanides or sulfides. They explode or detonate when heated in a confined space.
Toxicity	Any waste that can cause death, severe harm, or injuries and that can seriously affect health if ingested, aspirated, or brought in contact with the body.

Source: Adapted from Cantanhede, Alvaro. *Hazardous Waste Characterization and Classification Summary. Pan American Center for Sanitary Engineering and Environmental Sciences (CEPIS)*. Lima, Peru: CEPIS, 1994.

paper and construction, automotive, and printing and leather products industries. Although not required to report toxic releases to the EPA, agriculture and lawn-care industries are responsible for the release of toxic substances such as pesticides, toxic fertilizers, and some ignitable wastes (Rosenthal, 2004; Schaffer, 2001).

Biomedical and pharmaceutical wastes from hospitals, clinics, and laboratories also pose significant threats to humans and the environment if not managed properly. In the United States about 500,000 tons of regulated biomedical wastes are generated annually by approximately 380,000 generators such as clinics, hospitals, laboratories, and physicians' offices. The quantity of regulated biomedical waste is anticipated to grow significantly in the near future because of the growing trend in in-home health-care delivery coupled with the fact that most states now require biomedical wastes to be treated specially relative to other types of wastes (Gerrad, 1995). The military forces are a major culprit releasing significant amounts of highly toxic and radioactive substances into the environment.

Table 3.1b   Department of Transportation Hazardous Materials Classification System

Classification	Typical Hazardous Substance(s)
Flammable liquid	Gasoline, alcohol
Combustible liquid	Fuel oil
Flammable solid	Nitrocellulose (film), phosphorus
Oxidizer	Hydrogen peroxide, chromic acid
Organic peroxide	Urea peroxide
Corrosive	Bromine, hydrochloric acid
Flammable gas	Hydrogen, liquefied petroleum gas
Nonflammable gas	Chlorine, anhydrous ammonia
Irritants	Tear gas, monochloroacetone
Poison A	Hydroorganic acid, phosgene
Poison B	Cyanide, disinfectants
Etrologic agents	Polio virus, salmonella
Radioactive material	Uranium, hexafluoride
Explosives:	
Class A	Jet thrust unit
Class B	Torpedo
Class C	Signal flare, fireworks
Blasting agent	Blasting cap
Other Regulated Materials (ORMs)	
ORM A	Trichloroethylene, chloroform
ORM B	Calcium oxide, potassium fluoride
ORM C	Cotton, inflatable life rafts
ORM D	Small arms ammunition
ORM E	Ketone polychlorinated biphenyls (PCBs)

Source: US Office of Technology Assessment. Transportation of Hazardous Materials. Washington, DC: Government Printing Office, 1986.

It is a common practice to distinguish between inorganic and organic compounds. The former mostly consisting of heavy metals and trace minerals are examined first and the organic compounds are discussed thereafter with emphasis on their characteristics and adverse effects on health and the environment.

INORGANIC COMPOUNDS: TOXIC HEAVY METALS

The major inorganic elements including aluminum, arsenic, cadmium, chromium, copper, cyanide, iron, lead, mercury, nickel, silicates, tin, zinc, and their compounds are abundant in nature. The most dangerous inorganic elements include arsenic, cadmium, chromium, copper, lead, mercury, and zinc (Nebel and Wright, 2000; ATSDR, 1999). For centuries, some of these toxic metals have been relied upon as raw materials for industrial production. Advances in science and

Table 3.2 Sources and Types of Toxic Waste by Industry or Generators

Industry Waste and Source(s)	Type of Waste
Agriculture and Lawn Care	Pesticides, toxic fertilizers, ignitable wastes
Chemical Industry	Strong acids and bases, solvents and reactive wastes, and persistent organic pollutants (POPs)
Automobile, Aerospace, Automotive Repair, and Body Shops	Heavy metals, paint wastes (petroleum products). Used lead acid batteries' remnant solvents
Printing Industry	Heavy metal solutions, waste inks, spent solvents, electroplating waste inks, sludges with heavy metal
Metal Smelting and Refinery	Heavy metal wastes
Leather Products Manufacturing	Waste toluene and benzene
Paper Industry	Paint wastes/heavy metals, ignitable solvents, strong acids and bases
Construction Industry	Ignitable paint wastes, spent solvents, strong acids and bases
Hospitals, Clinics, and Laboratories	Biohazards, biological, medical, and pharmaceutical wastes
Military	Munitions and radioactive wastes

Source: Adapted from the US Environmental Protection Agency. *Understanding the Hazardous Waste Rules: A Handbook for Small Business.* Washington, DC: EPA, 1996 update.

technology and industrialization have contributed to a wide range of uses of these heavy metals and their trace elements. The mining of inorganic compounds and their use in industries have increased exponentially over the past two centuries (Nriagu, 1996). Through increased worldwide mining, processing, and consumption, the generation of wastes with a substantial component of heavy metals has skyrocketed (US Department of the Interior (DOI)/Bureau of Mines (BOM), 1993).

Since the Industrial Revolution, human releases of heavy metals have surpassed the systemic releases of these metals into the environment by nature. For instance, worldwide production and consumption of heavy metals such as mercury, lead, cadmium, beryllium, iron, and copper have increased substantially since World War II. As shown in table 3.3, the production of heavy metals has more than doubled, and the generation of metal-ore wastes has multiplied threefold in the United States since 1960. The advanced industrial societies consume the lion's share of the world's heavy metals, as well as generate the most wastes. Finding safe and sound mechanisms of disposing toxic wastes containing heavy metals remains a

Table 3.3 Metal Ore and Metal Waste Production in the United States, 1960–1990 (in Billion Short Tons and Percent)

Selected Year	Crude Ore	(%)	Waste	(%)	Type of Waste
1960	.421	45.0	.516	55.0	.938
1970	.586	37.5	.975	62.5	1.560
1980	.597	33.4	1.190	66.5	1.790
1991	.892	38.0	1.450	62.0	2.340

Note: Metals include bauxite, copper, gold, iron ore, lead, silver, titanium, zinc, etc.

Source: Adapted from the US Department of the Interior, Bureau of Mines. *Mining and Quarrying Trends in the Metal and Industrial Minerals Industries*. Washington, DC: DOI/BOM, 1993.

serious challenge in the United States. With the growing "Not in My Backyard" (NIMBY) opposition to new facility siting, the transfer of scrap metals and other types of toxic waste to less developed countries (LDCs) has increased significantly from the 1980s to the present (see Greenpeace, 1994; Adeola, 2000b). However, Third World NIMBY is growing strong as an increasing number of underdeveloped countries are resisting or putting a complete ban on hazardous waste importation. The Basel Convention has imposed some restrictions on the export of toxic waste from high-income countries to LDCs.

Heavy metals are not biodegradable; they remain toxic for as long as they exist. Most of these compounds also bioaccumulate as they work their ways through the food chain or trophic levels. They are used for many arrays of industrial, commercial, agricultural, and household purposes. Serious illnesses have been linked to both acute and chronic exposure to these elements and their compounds. For example, mercury is a known neurotoxin, arsenic may cause skin and other types of cancer, and lead is associated with neurological disorder and mental retardation, especially among children (Rice and Silbergeld, 1996).The National Academy of Science has reported that the neurological effects of methylmercury toxicity are most pronounced and damaging to children under 12 years of age and those contaminated in the womb (Raines, 2001). Cadmium is linked with high blood pressure (BP), heart disease, and lung and prostrate cancers (Tsoukala, 1998; WRI, UNEP, UNDP, and World Bank, 1998). Selected toxic heavy metals including their sources, properties, possible pathways, and health effects are discussed in the following sections. As shown in table 3.4, these elements are among the top 20 toxic substances commonly found in waste streams and especially in hazardous waste sites across the country.

Table 3.4 Top 20 Hazardous Substances and Substances Most Frequently Found in Completed Exposure Pathways (CEPs)

Substances	Hazard Rank	No. of NPL Sites	CEP Order
Arsenic*	1	147	3
Lead*	2	206	1
Mercury*	3	74	12
Vinyl Chloride	4	75	21
Benzene	5	116	5
Polychlorinated Biphenyls (PCBs)	6	96	8
Cadmium*	7	105	6
Benzo(a)pyrene	8	46	23
Polycyclic Aromatic Hydrocarbons (PAHs)	9	55	26
Benzo(b)fluoranthene	10	28	36
Chloroform	11	81	15
DDT, P'P'-	12	28	44
Aroclor 1260	13	13	83
Aroclor 1254	14	17	76
Trichloroethylene	15	239	2
Chromium (+6)*	16	102	7
Dibenz[a,h]anthracene	17	25	42
Dieldrine	18	19	56
Hexachlorobutadiene	19	—	—
DDE P,P'	20	37	44

Note: Items with an asterisk denote toxic heavy metals commonly found in waste streams.

Source(s): ATSDR/EPA Priority List for 1999; Substances Most Frequently Found in Completed Exposure Pathways (CEPs) at Hazardous Waste Sites.

ARSENIC

Arsenic is ranked number one every year among the top 20 most hazardous substances commonly found in waste streams in the United States (ATSDR, 2007) (see table 3.4). It is a naturally occurring heavy metal found in nature—present in soils, rocks, water, and plants. Given the fact that arsenic is a naturally occurring element in the Earth's crust, some level of exposure to this metal is inevitable. Civilizations have found a wide variety of uses for this heavy metal. Ancient Greek, Roman, Arabic, Peruvian, and Egyptian civilizations used the compounds of arsenic therapeutically, as poisons, and for other purposes. In contemporary society, arsenic compounds are employed in the production of insecticides, herbicides, fungicides and rodenticides; desiccants and defoliants used to facilitate the mechanical harvesting of cotton; and in wood treatments. Arsenic is also used in the glass industry and other metal-smelting operations. These industrial processes represent the sources of arsenic waste releases into the environment. Of course, arsenic is also released into the environment through the systemic process of weathering of rocks and due to surface run-offs. As noted by Wang and Rossman (1996: 221) and Berman (1980), arsenic may be released from the soil or from rocks into hot spring waters. Thus, this heavy metal is commonly found in drinking water, seawater, and food such as vegetables, grains, fruits, and seafood.

Clear scientific evidence has established the toxicity of arsenic. In higher concentrations, this metal is poisonous. It is a known carcinogen associated with various types of cancer including bladder, kidney, liver, and lung cancers. Exposure to higher concentrations of arsenic (> 400 μg/day) may cause death while exposure to lower levels (100–400 μg/day) may induce morbidity conditions such as abnormal heart functions, liver and kidney dysfunctions, gastrointestinal tract, nerves, and skin disorders (Wang and Rossman, 1996; ATSDR, 1998). Furthermore, nonallergic contact dermatitis and conjunctivitis have been found among workers exposed to arsenic-containing dusts. Other serious health problems associated with arsenic poisoning include birth defects, mental and physical impairments especially in children, loss of memory, and suppression of immune systems (Berman, 1980; Crawford, 1997).

Asbestos

Asbestos is a naturally occurring fibrous mineral of six different forms: (1) amosite, (2) chrysolite, (3) crocidolite, (4) termolite, (5) actinolite, and (6) anthophyllite. The most popular mineral type is white (chrysolite); however, other varieties may be blue (crocidolite), gray (anthophyllite), or brown (amosite). Asbestos fibers do not possess any typical odor or taste. They are resistant to heat and chemicals and as such, asbestos fibers have been used in a wide range of products including heat-resistant fabrics, insulation of electrical wiring, hot pipes and furnaces, building materials, and friction products. Asbestos is commonly used as an insulating and flame-retarding material, roofing material, floor covering, and in automobile brake shoes (ATSDR, 1995; Chapman, 1998). Incidentally, these diverse applications represent the major sources of asbestos waste generation and release into the environment. Approximately over 5 million metric tons of asbestos wastes are generated annually in the United States.

In its natural form, asbestos is generally not harmful. However, it becomes hazardous when improperly disposed off as waste and through the breakdown of its fibers over time. The wearing down of products made with asbestos releases ultrafine asbestos fibers, which become airborne and quite toxic upon exposure. Asbestos waste generally comes from mining, construction, automotive repair, and older building structures. Asbestos is very dangerous because of its fibrous nature and its nonbiodegradable property. Exposure to asbestos fibers has been linked to a serious health problem known as asbestosis—a noncancerous lung disease. Epidemiological studies have identified the health hazards of asbestos exposure as causing the following: leukemia, lung cancer and mesothelioma, cancers of the pleura, peritoneum, larynx, and pharynx, and oral cavity, esophagus, stomach, colon and rectum, and kidney, respectively (see Epstein et al., 1982; EPA, 1985; Chapman, 1998; ATSDR, 1995).

The federal government in the United States has taken several measures to protect citizens from exposure to asbestos. The EPA has banned new uses of asbestos as of July 12, 1989, and the agency has established regulations that require schools to inspect their buildings and remove asbestos. The agency has promulgated a national emission standard for asbestos under section 112 of the Clean Air Act, which established asbestos-disposal requirements for active and inactive waste sites. This regulation requires owners and operators of demolition and renovation projects to follow specific guidelines and procedures to prevent asbestos releases to the outside air, and further requires that demolition and renovation materials be controlled if the materials contain more than 1 percent asbestos by weight in a form that hand pressure can crumble, pulverize, or reduce to powder when dry (EPA, 1985: 4–34). These initiatives notwithstanding, thousands of homes across America still contain asbestos materials, posing substantial health risks to the residents. Furthermore, developing nations of the global South represent the paths of least resistance, where asbestos banned in the United States and other developed nations are sold openly in the market for building construction.

CADMIUM

Cadmium is also a naturally occurring metal found in the Earth's crust. It is silvery white, malleable, and with a bluish appearance; its odorless and tasteless characteristics make detecting its presence difficult without a chemical assay. Cadmium is rarely found in its pure form because it reacts with other elements in several compounds to form extremely toxic products. It is ubiquitous in the ecosystem—because it is found in the air, soil, water, sediments, and in the food and fiber system. Researchers have indicated that human activities such as mining, smelting, and refining of ores are directly linked to increased concentration of cadmium in the environment. Melting zinc, lead, and copper ores, fossil fuel combustion, incineration of waste, and production of steel are among the pathways through which cadmium wastes are released into the environment (EPA, 2006; Wang and Tian, 2004; Chedrese et al., 2006; Bernard, 2008). Populations living in proximity to industrial processing facilities, waste sites, and landfills are especially at risk of exposure to cadmium. Also at risk of exposure are workers in the mining, smelting, metal-plating, and e-waste processing industries. Clearly, soil and water in proximity to industrial facilities and landfills may contain higher concentrations of cadmium, which may pose greater risk to humans and other organisms.

Furthermore, cadmium is among the several hazardous substances found in e-waste. More specifically, it is used in rechargeable nickel-cadmium (NiCd) batteries, in the fluorescent layer of cathode ray tube (CRT) screens, in printer inks and toners, and in the printer drums of photocopying machines. It is also commonly used to electroplate steel, copper, iron, and brass to protect against oxidation and corrosion. Cadmium is employed to chemically stabilize some plastic materials. However, the percentages of traditional uses of cadmium for coatings, pigments, and stabilizers have gradually declined mostly due to environmental and health concerns (US Geological Survey, 2008).

This heavy metal has been determined to be highly toxic, persistent, with an estimated elimination half-life of 10–30 years, and it bioaccumulates in the human body, especially in vital organs such as the liver and the kidneys (Nawrot et al., 2008; Bernard, 2008). The US EPA and the Agency for Toxic Substances and Disease Registry (ATSDR) have recognized cadmium as a persistent and bioaccumulative pollutant included in the list of national priority chemicals. In 2006, 63 facilities reported generating about 953,000 pounds of cadmium waste, with 39.65 percent disposed on-site and 60.35 percent disposed off-site (EPA, 2006).

In the past two decades, increased environmental awareness has resulted in regulatory pressure to reduce or eliminate the use of cadmium in many developed countries. For instance, industrial releases of this element and other heavy metals are regulated and monitored by the EPA, ATSDR, and state environmental agencies across the United States. The European Union (EU) has been reviewing a proposal to ban, with few exemptions, NiCd batteries containing more than 0.002 percent cadmium, and to increase the fraction of spent industrial and portable rechargeable batteries collected. Despite the declining production of cadmium in developed countries, China and other developing countries have increased their cadmium production to meet the demands of the growing industry manufacturing NiCd batteries (USGS, 2006, 2008).

As mentioned, once released and dispersed in the environment, cadmium tends to persist in soil, water, and sediments for several decades. When absorbed by plants, it typically bioaccumulates in the food chain and eventually accumulates in the vital organs of the body. People who consume food crops grown on soils with high cadmium content, eat seafood harvested from cadmium-polluted water, smoke cigarettes excessively, or are employed in cadmium-processing factories are susceptible to cadmium toxicity (Bernard, 2008; Satarug and Moore, 2004). The natural and man-made sources of cadmium encompassing industrial releases and the application of phosphate and sewage sludge fertilizers to arable lands, may result in environmental contamination of soils, groundwaters, and plants that could yield unprecedented concentrations of cadmium in the food chain (Chedrese et al., 2006; Bernard, 2008). For nonsmokers, consumption of foods of plant etiology, such as cereal, grains, root tubers such as yam, cassava/tapioca, and potatoes, and vegetables, is considered a potential source of exposure to cadmium intake (Olsson et al., 2002; Amzal et al., 2009). Reports from different parts of the world reveal substantial variations in the daily dietary intake of cadmium, contingent upon dietary habits. While cadmium concentrations in staple foods vary considerably, the most average concentrations are found in seafood (e.g., fish and marine mollusks), cereals, grains, leafy vegetables, root tubers, and mushrooms (Chedrese et al., 2006: 28).

In light of cadmium's properties of toxicity, bioaccumulation, persistence, and ubiquity in the environment, this element poses a serious, present, and continuous threat to human health and those of other species in the ecosystem. Public awareness and concern about the adverse health effects of cadmium toxicity evolved during the post–World-War II epidemic of "Itai-Itai" (Ouch-Ouch) disease in Japan. The victims of "Itai-Itai" disease were exposed to high cadmium toxicity through the consumption of rice and other food crops produced by farmers who used cadmium-contaminated run-off water from the Kamioka mine along the Jinzu River basin to irrigate their rice paddies and other field crops.

Over time, the local women presented cases of severe skeletal demineralization, osteomyelitis, and osteoporosis. Postmenopausal women who had several children were particularly vulnerable to the disease. As revealed by the "Itai-Itai" disease, in extreme cases of cadmium toxicity, aching of bones and joints culminating in brittleness, fracture, and deformity of bones are among the adverse health outcomes. In addition to bone or skeletal destruction, cadmium toxicity has also been associated with liver and kidney dysfunction, high BP, heart disease, depressed immune system, leukemia, and cancers of the lung, pancreas, kidneys, bladder, breast, and prostate (Chedrese et al., 2006: 28). Recently, researchers have revealed a novel concept of cadmium acting as an endocrine-disrupter chemical.

The health effects described above are chronic in nature, manifesting due to a prolonged exposure and bioconcentration of cadmium in the body tissues. Acute poisoning of cadmium is also a serious health concern. Occupational exposure or industrial accident can trigger acute cadmium poisoning. Among the symptoms of acute exposure to high concentrations of cadmium fumes are flulike characteristics such as weakness of the body, fever, profuse sweating, chills, joint and muscular pains, and death.

LEAD

Similar to asbestos, lead is ubiquitous in the environment and poses a serious problem, especially among the working classes and the lower-income homes in the central cities in the United States. Lead is considered as one of the metals of antiquity that has been used by all known human civilizations. For example, ancient Greek and Romans used lead for a wide variety of purposes, and lead toxicity was recognized in antiquity (Berman, 1980). Some scholars have speculated that lead toxicity might have contributed to the fall of the Roman Empire (see Berman, 1980; Drotman, 1985). It is a highly toxic metal that poses major health risks to exposed populations. It is a naturally occurring element in the Earth's crust; this metal does not display any characteristic taste or smell and will not burn or dissolve in water. However, lead can bind with other chemicals to form lead salts or compounds. Some man-made lead compounds can burn, for example, organic lead compounds in gasoline. With mounting epidemiological evidence based on studies carried out in the past three decades, many public health authorities in the United States have concluded that there may be no acceptable level of lead exposure (CDC, 1991).

Annually, more than 9 million metric tons of lead waste are generated in the United States. The EPA has found lead in at least 1,026 (about 70 percent) of the 1,467 waste sites currently placed on the National Priority List (NPL). This finding reflects the fact that lead is widely dispersed in the environment. The use of lead in industry has grown exponentially since time immemorial. In contemporary society, lead has been used for several purposes including the manufacture of batteries, glass, ammunition, a wide variety of metal products (e.g., sheet lead, solder, pipes, and brass and bronze products), medical equipment, paints, ceramics, scientific equipment, and military equipment. In the past, lead compounds such as tetraethyl lead and tetramethyl lead have been used as gasoline additives to increase octane rating in the United States. The use of these chemicals was discontinued in the 1980s, and lead was banned for use in gasoline effective January 1, 1996.

According to the ATSDR (1997), most of the lead used by industry is derived from mined ores or from recycled scrap metals or batteries. Human activities such as the use of leaded gasoline and lead-based paints have released lead and substances containing lead throughout the environment. Substantial amounts of lead can be encountered in the air, soil, drinking water, rivers and lakes, and in most building structures. As a result of the ban on leaded gasoline, the release of lead into air through automobile exhaust pipes has declined significantly in recent years. The majority of the lead in inner city soils is from older homes with lead-based paints.

There are many sources of exposure to lead and lead products including residential proximity to hazardous waste sites, living in older homes, exposure to e-wastes, consumption of foods contaminated by lead, drinking water from lead pipes, and using health-care products that contain lead. Substances such as folk remedies, certain health foods, cosmetics, and "moonshine whiskey" are sources of lead exposure that are difficult to monitor or control. Occupational exposure is common among people employed in mines, soldering shops, plumbing and pipe fitting, lead smelting and refining industries, brass/bronze foundries, battery-manufacturing plants, and lead-compound-manufacturing facilities. Construction workers and employees of municipal waste incinerators are also prone to exposure to lead (Drotman, 1985). Also at risk of lead exposure are people working at e-waste recycling facilities or living in proximity to e-waste dumping sites.

Exposure to lead poses a variety of health hazards. Even in small doses, neurotoxicity, intelligence quotient (IQ) deficits, learning

disorder, and attention span disorder and hyperactivity, especially
among children, are possible health effects of lead exposure (Rice
and Silbergeld, 1996; ATSDR, 1997; Drotman, 1985). At high
concentrations (> 15 µg/dL), lead may cause significant damage to
the brain and kidneys in adults and children as well. Other health
problems include cancer, hearing loss, and chronic neurobehavioral
dysfunctions. It has been associated with miscarriage among pregnant women. Lead poisoning may also result in anemia, increased
BP, hypertension, and associated cardiovascular disease (ATSDR,
1997; Epstein et al., 1982; Drotman, 1985; Chapman, 1998; WRI,
UNEP, UNDP, and World Bank, 1998). Although there is no sub-
group, either by race or socioeconomic status, immune to lead toxicity, minority groups such as African Americans, Mexican Americans,
Puerto-Rican Americans, and American Indians and their children
are at higher risk of lead exposure and associated health diminution.
According to the ATSDR, approximately 17 percent of children in the
United States are at risk of lead poisoning; among minority groups,
about 46 percent of African American children were estimated to be
at risk of lead toxicity.[1] African Americans are disproportionately at
risk of occupational and residential exposure to lead and other xenobiotics as a result of a combination of factors including poverty, historical housing, and occupational discrimination, and the patterns
of hazardous facilities placement in low-income powerless neighborhoods (Adeola, 1994).

MERCURY

Similar to cadmium or lead, mercury is a naturally occurring heavy
metal that exists in many forms and is present throughout the environment. According to the United Nations Environmental Program
(UNEP), the levels of mercury in the environment have increased
phenomenally since the onset of the Industrial Revolution. Currently,
mercury and its compounds are found in various environmental
media and food (particularly seafood) around the globe. High levels
of mercury adversely affect the health and well-being of humans and
wildlife. The three basic chemical forms of mercury are elemental or
metallic mercury, inorganic mercury, and organic mercury (such as
alkylmercury, phenylmercury, methylmercury, and methoxyethylmer-
cury (Hamada and Osame, 1996; ATSDR, 1999). As a pure form of
mercury, elemental mercury is the common liquid used in thermometers, barometers, and some electrical switches. Inorganic mercury
compounds are produced when elements such as oxygen, chlorine,

sulfur, etc., combine with elemental mercury to form mercuric oxide, mercuric chloride, mercuric sulfide, etc. (UNEP, 2002).

Most mercury air emissions are in the gaseous elemental mercury state, which often travels far and wide across the globe, even to regions so distant from the original source of emission. Interaction of metallic mercury with carbon also yields organic mercury compounds (see ATSDR, 1999; Allchin, 1999; Smith and Smith, 1975). As noted by Hamada and Osame (1996), each form of mercury produces toxic effects in humans and each form can be transformed to the other form. For instance, some microorganisms and other natural factors can transform each form of mercury to another form becoming more lethal to humans and other life forms. Methylmercury produced by these agents generally bioaccumulates within the food chain, especially becoming more concentrated in fish and marine mammals as found in the case of Minamata, Japan (see Smith and Smith, 1975).

Elemental mercury has many diverse uses including for gold extraction and the manufacture of thermometers, barometers, batteries, and electrical switches as mentioned earlier. Inorganic mercury compounds have been used as fungicides, and mercuric chloride and iodide are major ingredients of some cosmetics such as skin-bleaching creams. Mercuric chloride has also been used as a topical antiseptic agent in some medicated soaps in the past, and it has been used in medicinal products such as laxatives, in warming remedies, and in teething powders for children. Thus, mercury wastes are generated in numerous ways—including private household, industrial, commercial, medical, and laboratory sources. In most part, human activities such as mining, smelting, industrial processes, and burning of fossil fuels have substantially elevated the amount of mercury released into the environment.

According to the ATSDR (1999), about 80 percent of the mercury released into the environment through human activities is metallic mercury. The principal sources of metallic mercury emission include solid waste incineration, mining, smelting, and burning of fossil fuels. Approximately 15 percent of the total mercury in the environment is released into the soil via fertilizers, fungicides, and municipal solid wastes, especially containing discarded batteries, thermometers, fluorescent light bulbs, and electrical switches. Industrial waste water accounts for about 5 percent of mercury in the environment. In the United States, mercury has been found in most hazardous waste sites, especially those placed on the NPL—a list of the most dangerous hazardous waste sites in the United States compiled by the EPA, using the hazards ranking system (HRS).[2]

The problems of bioaccumulation and biomagnification of methylmercury within the food chain have been studied extensively. For instance, several cases of mercury-poisoning disasters include Minamata, Japan (where waste containing methylmercury was discharged by Chisso Corporation's chemical plant into the sea of Minamata). Other similar cases include Ontario, Canada; Pakistan; Guatemala; and Iraq (where organomercury-treated seed grains were consumed by humans (see Hamada and Osame, 1996; Allchin, 1999; Smith and Smith, 1975). The etiology of methylmercury poisoning in Japan was the reckless discharge of methylmercury compounds into the sea and the rivers of the fishing communities in Minamata, Japan. This episode was described by Smith and Smith (1975: 26) thus:

> Undoubtedly, the chemical company called Chisso contaminated the resource dependent communities in Minamata Bay area by poisoning the fishing waters, aquatic food chain, and a large number of the inhabitants. The company discharged methylmercury through waste pipes into Minamata Bay and turned the fishing waters into a sludge dump which destroyed the entire habitat, killing thousands of people including their culture and heritage.

Among the symptoms of methylmercury poisoning (also known as Minamata disease) are cerebellar ataxia, dysarthria, concentric constriction of the visual field, sensory impairment, and hearing impairment as illustrated in figure 3.1. Other symptoms include involuntary movements, nerve dysfunctions, impaired brain development, and other abnormalities. Elemental mercury toxicity manifestations include mental changes (including irritability, excitability, insomnia, and problems of concentrations and hallucinations), salivation, gingivitis, stomach upset, and contact dermatitis in cases of skin exposure to mercury (see Hamada and Osame, 1996: 341–2; Smith and Smith, 1975; ATSDR, 1999).

While elemental mercury produces acute toxic effects, methylmercury or organic mercury bioaccumulates and causes chronic (long-term) effects. Most people with severe symptoms of Minamata disease died within two to six weeks after the onset of the disease (Funabashi, 2006). The congenital aspect of the disease has been noted in the early 1960s. Many children received high doses of methylmercury while still in their mother's womb. Consequently, a significant number of children born in the contaminated villages suffered profound neurological disorders including deafness, blindness, cerebral palsy, and mental retardation (see figure 3.1).

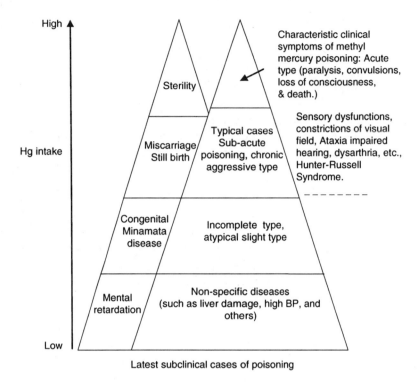

High

Hg intake

Low

Sterility

Characteristic clinical
symptoms of methyl
mercury poisoning: Acute
type (paralysis, convulsions,
loss of consciousness,
& death.)

Miscarriage
Still birth

Typical cases
Sub-acute
poisoning, chronic
aggressive type

Sensory dysfunctions,
constrictions of visual
field, Ataxia impaired
hearing, dysarthria, etc.,
Hunter-Russell
Syndrome.

Congenital
Minamata
disease

Incomplete type,
atypical slight type

Mental
retardation

Non-specific diseases
(such as liver damage, high BP, and
others)

Latest subclinical cases of poisoning

Figure 3.1 Hierarchy of Mercury Toxicity and a Wide Range of Symptoms of Minamata Disease.

Over the years, many people have died, afflicted with physical deformities, or have endured physical and emotional pains and poor quality of life as they struggled with Minamata disease. More than 3,000 people have been officially recognized in Japan as legitimately having "Minamata disease" and several thousands more are making their cases. Official certification or determination is required for victims to qualify to receive compensation, which has been delayed in protracted legal battles spanning more than 50 years. In summary, the symptoms of Minamata disease are:

- Sensory impairment in the four extremities—loss of sensation in the hands and feet;
- Ataxia—difficulty in coordinating movement of limbs or hands and feet;
- Narrowing or constriction of the visual field;
- Hearing impairment;

- Impairment of faculties for maintaining balance;
- Speech impediments;
- Disorders of ocular movement;
- Paralysis, convulsions, and loss of consciousness; and
- Congenital disorders.

In Iraq, two instances of methyl- and ethylmercury poisonings involving the consumption of seed grain that was treated with fungicides containing these compounds have been reported. Imported mercury-treated seed grains were delivered after the planting season and were later used as grain processed into flour for human consumption. The first incidences caused by ethylmercury occurred in 1956 and between 1959 and 1960, and approximately 1,000 people were affected. The second episode was caused by methylmercury and it happened in 1972. An estimated 6,500 thousand people were admitted to the hospital with 459 fatalities. In contrast to the chronic exposures and prolonged health effects in Japan, the epidemic of methylmercury toxicity in Iraq was acute in nature, albeit with a high magnitude of toxicity. The tragedy occurred partly because the bags of the fungicide-treated grains were labeled in English and not in Arabic, which was the language understood by many peasants in the rural areas where the seeds were delivered. Since many of the people exposed to methylmercury toxicity in this manner were located in remote small villages (and some were nomads), the actual number of casualties due to mercury-tainted seed grains is unknown.

In the United States, mercury is found at approximately 50 percent of all the Superfund sites, and mercury contamination is the most common cause of pollution of rivers and lakes across the country. About 45 states have issued warnings about eating locally caught freshwater fish. The Food and Drug Administration (FDA) and the EPA released a consumer advisory in 2004 that recommends that young children, pregnant women or women that may become pregnant, and nursing mothers should avoid eating fish and certain seafood from the ocean, especially those that may contain high levels of methylmercury, such as king mackerel, shark, swordfish, and tilefish (EPA/FDA, 2004).[3] A study by the US National Institutes of Health (NIH) in 2005 estimated that 1 woman in 12 in the country has more mercury in her blood than the 5.8 µg/l threshold considered safe by the EPA, and that between 300,000 and 600,000 of the 4 million children born each year in the United States have been exposed in the womb to mercury levels that may lead to diminished intelligence or developmental dysfunctions (cf. Cunningham et al., 2007: 345).

THE ORGANIC COMPOUNDS

Organic compounds are produced in nature through the combination of carbon, hydrogen, and nitrogen molecules. Among the important properties of organic compounds are specific gravity (the degree to which they sink or float in water), solubility, volatility, adsorption (or the tendency to stick to the surface of other matter), and degradation through decomposition. The bonds of natural hydrocarbons are easily broken through decomposition as they are recycled. Advances in organic chemistry, however, have led to the artificial production of synthetic organic compounds that are more durable, lipophilic, versatile, and toxic than their natural counterparts. Synthetic organic compounds are derived from a petroleum base. These petrochemical organic compounds are used in the manufacture of a wide array of modern products including plastics, cooking utensils, synthetic fabrics, synthetic rubber, paintlike coatings, solvents, pesticides, toxic wood preservatives, and numerous other products. As mentioned earlier, synthetic organic chemicals are in most part, the products of the twentieth-century chemical revolution. Two broad categories of organic compounds that pose serious health risks to humans and the environment are POPs including organochlorine products and non-persistent organic pollutants (non-POPs). In general, synthetic organic compounds are toxic to humans, decomposer organisms, and the environment. POPs are not biodegradable and generally bioaccumulate in the environment as compared with natural organic compounds (Nebel and Wright, 2000:478; Thornton, 2000; Epstein et. al., 1982).

Table 3.5 presents a list of anthropogenic organic compounds commonly found in waste streams across the United States and their established adverse health effects. The 12 POPs asterisked are of the most concern to the international community. Some countries have restricted or completely banned these chemicals. Even though the use of POPs is restricted or banned in developed nations, the United States and other advanced industrial nations continue to manufacture them for export to Third World countries.

POPs are defined as a group of synthetic organic compounds that, to a varying extent, are resistant to degradation, soluble in lipids but not in water, semivolatile, toxic, bioaccumulative, and capable of migrating long distances from the point of original release. These chemical compounds have been found in remote regions of the world

Table 3.5 Toxic Synthetic Organic Compounds Commonly Present in Chemical Wastes and Their Health Effects

Toxic Chemical	Established Adverse Health Effects
Aldrin*	Dizziness, nausea, malaise, and liver and biliary cancers
Benzene+	Mutations, cancers, birth defects, and stillbirths
Carbon tetrachloride+	Cancers, birth defects, stillbirths, neurotoxicity, hepatotoxicity, and kidney diseases
Chlordane*	Cancers (tests remain inconclusive)
Chloroform+	Cancers, birth defects, embryo toxicity, and hepatotoxicity
Chloroethylene (vinyl chloride)+	Mutations, cancers, nervous disorders, liver disease, and lung disorders
Chlorotoluene+	Mutations, cancers
Dichlorobenzene (DCB)+	Mutations, nervous disorders, liver disease, and kidney disease
Dichloroethylene+	Mutations, cancers, birth defects, stillbirths, nervous disorders, and liver and kidney disorders
Dieldrin*	Liver and biliary cancers
Endrin*	Cancers
Furfural+	Mutations, nervous disorders
Heptachlor*+	Mutations, cancers, stillbirths, birth defects, and liver disease
Hexachlorobenzene (HCB)*	Mutations, cancers, birth defects, fetal and embryo toxicity, nervous disorders, liver disease, photosensitive skin lesions, and hyperpigmentation
Mirex*	Acute toxicity, possible cancers
Polychlorinated Biphenyls (PCBs)*+	Mutations, cancers, birth defects, fetal and embryo toxicity, neurological disorders, and liver disease
Polychlorinated Dibenzo-Para-Dioxins and Furans*	Peripheral neuropathies, fatigue, depression, hepatitis, liver disease, abnormal enzyme levels, hepatotoxicity, chloracne, embryo toxicity, and gastric lesions
Tetrachloroethylene+	Cancers, nervous disorders, liver disease, and kidney dysfunctions
Toluene+	Mutations, birth defects, stillbirths, and nervous disorders
Toxaphene*	Cancers, chromosome aberrations, and liver and kidney dysfunctions.
Trichloroethylene+	Mutations, cancers, nervous disorders, and liver disease
1-1'-(2,2,2-Trichloroethylidene) (4-chlorobenzene) (DDT)*	bis Cancer of liver, immune system suppression
Xylene+	Birth defects, stillbirths, and nervous disorders

Note: Asterisks designate those compounds commonly present in hazardous waste and + designates POPs.

Source: Adapted from Epstein, S. S., L. O. Brown, and C. Pope. *Hazardous Waste in America*. San Francisco, CA: Sierra Club Books, 1982, pp. 415–427 and http://irptc.unep.ch/pops/indxhtms/asses0-html.

where they have never been manufactured or used. Two basic groups of POPs are polycyclic aromatic hydrocarbons (PAHs) and some halogenated hydrocarbons, which include several organochlorines. The organochlorines are the most persistent compounds with wide production, use, and toxic releases.

As McGinn (2000: 80) points out, some of these synthetic organic compounds were in fact considered modern miracles that have helped humans gain substantial control over nature, especially by increasing the levels of food production and health standards through the control of pests, insect-borne diseases, and soil fertility. Paul Muller received the Nobel Prize for developing dichlorodiphenyltrichloroethane (DDT) in 1948. However, numerous previously unknown adverse health effects of DDT and other POPs have been identified since 1960s. As a legacy of the anthropogenic organic chemical revolution of the twentieth century, a toxic brew of thousands of chemicals is now ubiquitous in every biome on the planet. The outcome is a litany of environmental problems that have permeated the mass media since the 1960s, including: DDT and the decline of bald eagles; toxic waste at Love Canal and numerous communities; cancers; PCBs in polar bear tissue; groundwater contamination; mercury and dioxin contamination and fish-kills; embryo toxicity; hormone dysfunctions; breast-milk contamination; vital organ dysfunctions, and many new adverse health conditions (see Thornton, 2000: 2; Thomas et al., 2001).

The POPs that are well recognized for their persistence, bioaccumulation, and biomagnification within the food web are organochlorines or organohalogens. This group includes dioxins and furans (produced and released through the incineration of industrial wastes, combustion, and industrial processes), PCBs, hexachlorobenzene (HCB), mirex, toxaphene, heptachlor, chlordane, and DDT (see table 3.5 for a complete list). PCBs, dioxins, and furans are among the several modern industrial products or by-products that have been found to be extremely harmful to humans and the environment. Halogenated compounds, especially organochlorines, are used by the chemical industry in the manufacture of a wide variety of products including polyvinyl chloride (PVC), solvents, pesticides (used in agriculture and municipalities), and specialty chemical and pharmaceutical products. Table 3.5 also presents other synthetic organic chemicals most commonly found in hazardous waste sites across the United States and their known health hazards. A more detailed discussion of POPs and their chemical properties and adverse health effects on humans, other organisms, and the environment is the focus of chapter 5.

Non-POPs are nonchlorinated hydrocarbons, which are generally soluble, volatile, poorly adsorbed, but rapidly degradable. They are mostly used as solvents (e.g., xylene, acetone, ethyl benzene, and methyl isobutyl ketone), in paints and varnishes, as engine and machinery cleaners, in the manufacture of rubber, for fuels and gasoline, and in several other industrial applications (Tammemagi, 1999: 71). Even with their relative nonpersistence, these synthetic chemicals pose a number of serious health concerns ranging from birth defects and cancers, to stillbirths and liver and kidney problems. Industries are largely responsible for the production and improper releases of synthetic organic compounds including the POPs and non-POPs. The cases of Love Canal, New York; Woburn, Massachusetts; Agriculture Street, New Orleans; Bhopal, India; and Minamata, Japan, reveal the economic, environmental, health, and psychosocial dimensions of community toxic contamination. Most of the chemicals buried at the Love Canal in particular are of the POP group. In addition to toxic organic compounds discussed thus far, nuclear waste represents a serious dread posing significant threat to human health and the environment. The following section addresses the varieties of nuclear waste and the risks associated with improper or accidental releases of such waste.

NUCLEAR OR RADIOACTIVE WASTE

Nuclear or radioactive waste refers to those materials that contain radionuclides or that have been contaminated by radionuclides, but are considered to be no longer useful. According to the International Atomic Energy Agency (IAEA), radioactive waste is any material for which no use is foreseen and which contains radionuclides at concentrations greater than the values deemed acceptable by the competent authority in materials suitable for use not subject to control.[4] The categories of nuclear waste distinguished by the Nuclear Waste Policy Act of 1982 include "high-level radioactive waste" (HLRW), transuranic waste (i.e., material considered as waste that contains radionuclides with atomic number greater than that of uranium), and "low-level radioactive waste" (LLRW), respectively.

Radioactive wastes can exist in solid, gaseous, liquid, or sludge states and their level of radioactivity can vary (Saling and Fentiman, 2001; EPA, 2007). For instance, the solid nuclear waste is derived from the mining and milling of uranium and thorium ores, sludges in storage tanks containing waste solutions, and from contaminated equipment and structures. The liquid radioactive waste comes in most part from spent fuel-reprocessing plants, for example, the aqueous waste from the first-cycle extraction system in irradiated fuel processing. And, the gaseous radioactive waste comes from the gaseous releases of reprocessing plants and nuclear power plants. This type of nuclear waste typically contains airborne radionuclides.

Nuclear wastes may be classified based on their origin (e.g., commercial energy waste, defense waste, etc.), the content of the waste (i.e., transuranic waste or spent nuclear fuel), physical properties of the waste, the intensity or level of radioactivity, such as high-level and low-level radioactive wastes, and the radioactive period or half-life, such as long-lived waste if the half-life is greater than 30 years and short-lived waste if less than 30 years (Comte and Flüry-Herard, 2005–2006).[5] Radiation is the amount of energy moved as particles or waves move through space or from one object or body to another (Shrader-Frechette, 1993). For the present purpose, classification by level or intensity of radioactivity and physical state of the waste is employed.

CLASSIFICATION BY TYPE, LEVEL OF RADIOACTIVITY, AND PHYSICAL STATES

(a) High-Level Radioactive Waste (HLRW)

HLRW is defined as the highly radioactive material produced through the processing of spent nuclear fuel, including liquid waste produced directly in reprocessing, and any solid material derived from such liquid waste that contains fission products in sufficient concentrations. The amount of plutonium and other isotopes left in the solutions is small and the residue consists of fission products. HLRW is produced in nuclear power plants by splitting, or fission, of uranium atoms in a controlled nuclear reaction (Tammemagi, 1999: 69; Saling and Fentiman, 2001; Tomain, 2004). This type of waste bears the major component of radioactivity (over 95 percent); consequently, it is the seat of considerable energy release, this remaining significant on a scale of several centuries. Consequently, the major method of disposal being considered is underground storage.

In the United States, the Yucca Mountain in Nevada has been designated as the appropriate site for permanent deep burial of HLRW based on the recommendation of the Bush administration's Department of Energy (DOE) in 2002. This recommendation and the decision to use Yucca Mountain as a dumpsite for all HLRW in the United States have drawn stiff opposition from the Native Americans who own the land, the residents of the state of Nevada, and over 50 national and 700 regional, state, and local grassroots environmental organizations.[6] As of the time when this was written, it remains doubtful whether the proposed site would ever materialize, especially under the Obama administration.

(b) Low-Level Radioactive Waste (LLRW)

LLRW refers to radioactive material that is not high-level radioactive waste, spent nuclear fuel, transuranic waste, or by-product material as defined in section 11e(2) of the Atomic Energy Act of 1954. It contains relatively little radioactivity mostly with no transuranic elements. About 99 percent of the radioactivity and half the volume in LLRW are produced by nuclear power plants, mostly the ion-exchange resins that filter radioactivity from reactor-cooling water. Other sources are industrial, medical, and research applications (Gerrard, 1995: 32). In recent years, the volume of civilian LLRW generated has been declining substantially due largely to higher disposal surcharges.

The method of disposal of LLRW is typically by specially prepared landfills other than municipal landfills. Precautions are taken to prevent inflow of water into the waste and leaking of contaminated water out of the site. This is done through an understanding of the topography of the site and geomorphology of the surroundings. The use of lining materials in the construction of the landfills and covering wastes to be disposed with impermeable materials prevent seepage of contaminated materials into underground water supply and the environment. Disposal of LLRW is not considered as problematic as any other type of radioactive waste and, as such, has generated less controversy, especially relative to HLRW.

(c) Transuranic waste (TRUW)

TRUW material refers to radioactive waste such as plutonium (Plutonium[239]), which has an atomic number greater than that of uranium (Uranium[235]). It is defined as waste contaminated with alpha-emitting transuranic radioisotopes that have a half-life of greater than 20 years and a concentration level above 100 nanocuries per gram (SRNS, 2008). Even though most TRUW contains a relatively low level of radioactivity, the fact that it is persistent or long-lived and highly toxic raises environmental and health concerns. According to the US DOE (2009), TRUW contains man-made elements heavier than uranium, such as plutonium as noted above, which explains the name "trans" or "beyond uranium." It

Figure 3.2 Contact-Handled TRU Waste Being Processed in an Airlock Room.
Source: US Department of Energy (DOE), Oak Ridge Environmental Management Program, April, 2009.

Figure 3.3 A Shipment of TRU Waste Leaving Savannah River Site (SRS) to Waste Isolation Pilot Plant (WIPP).
Source: US DOE; http://www.srs.gov/general/news/factsheets/truw.pdf (accessed January 23, 2010).

is generally connected with the human manipulation of fissionable material dating back to the Manhattan Project through the present time. How TRUW is handled is basically determined by its composition. Higher-energy radioactive TRUW is called "remote-handled," which means it must be processed by remote control equipment in a special facility known as "hot cells." People processing remote-handled waste are protected by barriers such as thick concrete walls and leaded-glass viewing windows. TRUW with low-energy radioactive content is considered as "contact-handled," which means that it can be handled without remote equipment, albeit the workers never actually touch the waste without using protective barriers such as special clothing, gloves, and equipment as illustrated in figures 3.2 and 3.3. The waste isolation pilot plant (WIPP) near Carlsbad, New Mexico, provides permanent isolation and disposal of TRUW in underground salt caverns.[7]

MAJOR SOURCES OF NUCLEAR WASTES

People have generated nuclear wastes as a by-product of various endeavors since the discovery of radioactivity in 1896 by Antoine Henri Becquerel (Jablonski, 2009). Nuclear waste is generated from several sources including mining and milling processes, nuclear power generation, medical and industrial processes, defense or military weapons production and testing, and scientific research (Andrews, 2006). All items and equipment used during these industrial, medical, and research activities, such as glassware, protective clothing, rags, tools, machinery, and plastic bags are contaminated with radioactive material and must be treated as radioactive waste posing serious danger to human health if not properly disposed (EPA, 2007b).[8] Unlike hazardous wastes, which are generated from every sector of the economy, most anthropogenic nuclear wastes in the United States are generated using only limited enterprises or sources—including commercial nuclear power plants for the generation of electricity and the production of nuclear weapons by the defense industry (Gerrard, 1994: 25). About 20 percent of electricity supply in the United States is derived from nuclear power reactors; thus, radioactive wastes are mostly generated by commercial nuclear reactors in the course of electricity production (EPA, 2007b). Table 3.6 summarizes the major sources of nuclear waste types in the United States. Clearly, defense-related radioactive wastes are the most pervasive and commercial nuclear cycle operations for electricity generation and they are directly linked to spent fuel, HLRW, and LLRW, respectively.

Table 3.6   Major Sources of Nuclear Waste Types

Sources	Types of Nuclear Wastes			
	Spent Nuclear Fuel	HLRW	TRU	LLRW
Commercial nuclear fuel cycle				
Operations for electricity	+	+		+
Decontamination and decommissioning of fuel cycle		+	+	
Defense related	+	+	+	+
Industrial users		+	+	+
Research, medical, universities	+		+	+

Source: Adapted from Saling, James H. and Andeen W. Fentiman. *Radioactive Waste Management*. 2nd ed. Boca Raton, FL: CRC Press, 2001.

The locations where the United States manufactured nuclear weapons are generally referred to as nuclear weapons complex (NWC). It includes 14 major facilities in 13 states. According to Gerrard (1994: 35), the NWC has left behind a legacy of environmental horror because at virtually every facility within the NWC, the groundwater is severely contaminated with radionuclides, and surface waters, sediments, and the soil are extensively polluted. Clearly, these pose serious health threats to humans and other organisms in the areas. More than half of NWC sites have been placed on the EPA's NPL. The major problem with nuclear waste is the fact that upon release, radioactivity may persist for many thousands of years posing significant health threat to humans and other life forms in the environment across wide geographical areas.

The intensity of radioactivity is generally measured by the curie, defined as the quantity of a radioactive isotope that decays at the rate of 3.7 x 10^{10} disintegrations per second. Basically, the greater the intensity of radioactivity, the greater the damage it can do to the exposed body. Shrader-Frechette (1993: 14) explains the short-term and long-term health effects of exposure to radiation. The acute or short-term effects of severe exposure typically include nausea, vomiting, dizziness, loss of consciousness, headache, etc. Among the long-term health effects of chronic exposure to radiation are birth defects, cancer, death, genetic defects, and reproductive failure. It was further indicated that there is no threshold for increased risk due to exposure to even small amounts of radiation. The fact that radiation effects are cumulative implies that successive exposures increase one's risk of serious bodily harm.

Just as Shrader-Frechette (1993) have noted, Tammemagi (1999: 69) also notes that the major adverse effect of nuclear waste radiation on humans is cancer, which often may not become evident until several years after exposure. Radioactive compounds can cause serious health problems even from a distance as well as through direct contact and ingestion. The Chernobyl nuclear disaster in the former Soviet Union (including Belarus, Ukraine, and the Russian Federation) vividly illustrates the adverse health effects of nuclear radiation over a wide geographical area. Genetic defects in the children of parents exposed to nuclear waste radiation and mental retardation in the children of mothers exposed to radiation during pregnancy have been reported in the literature (EPA, 2007; Medvedev, 1990). Also reported are acute radiation syndrome mortality, cancer mortality and morbidity, thyroid cancer in children, leukemia, cataracts among exposed children, and psychosocial problems including symptoms of stress, anxiety, depression, and medically undiagnosed or unexplained physical or physiological malaise (see The Chernobyl Forum, 2003–2005). Moreover, individuals within the affected populations have had to struggle with the stigma of being labeled as "sufferers," which later became known colloquially as "Chernobyl victims." This stigma has an adverse effect on the individuals' self-concept as it readily encouraged them to think of themselves as fatalistically invalid, helpless, weak, and needing extensive government benefits to survive.[9]

SUMMARY

This chapter offers some insights into the classification and identification of hazardous wastes and their deleterious health consequences on humans and the environment. It is essential to understand toxic wastes by sources, types, chemical composition, and associated risks. Several classification schemes—by industry, chemical composition, inorganic and organic types, radioactive waste types, and persistence in the environment have been discussed in this chapter. Furthermore, specific taxonomy of toxic and hazardous wastes and their adverse health effects on populations and the ecosystem have been presented. While quite extensive, the classification presented in this chapter is by no means exhaustive, partly constrained by page limitation among other considerations. There is a need for a universally accepted classification of hazardous waste to avoid ambiguities in the management and transportation of such waste from cradle to grave within and across nations. Clearly, more stringent environmental regulations of transboundary movement of toxic waste, especially from more affluent nations to underdeveloped societies of the global South, are imperative. Also imperative is finding a suitable permanent site for securing HLRW and TRUW.

Part II

Electronic Waste, Persistent Organic Pollutants, and Health Hazards

High-tech and chemical revolutions are two important aspects of globalization touching different regions of the biosphere. Therefore, the topics of electronic waste (e-waste) and persistent organic pollutants (POPs) naturally belong to the same segment of a book such as this. On the one hand, high tech represents an important component of the engine propelling globalization and economic development, but on the other hand, the e-waste generated now represents the dark side, often overlooked. There are numerous adverse impacts of highly toxic materials due to e-waste on human health and the environment. Similarly, the persistent organic compounds first touted as "modern miracles" have turned into "nightmares" as they spread across the globe, even into regions where they have never been produced or used.

Chapter 4 explores the problems of e-waste in great detail. The chapter identifies the sources of e-waste, their rates of growth, proliferation, and transnational movements. The patterns of flow of both e-waste and illicit POPs are also presented with excellent illustrations. Existing measures for controlling international flow of e-waste, especially from advanced industrialized nations to less developed countries (LDCs) are presented. Chapter 5 covers various aspects of synthetic POPs with emphasis on their sources, characteristics, potential or actual exposure pathways, and health effects on humans and wildlife. The salience of precautionary principles and the Stockholm Convention on POPs are addressed. Also, the regulatory measures for both e-waste and POPs within the United States and the international community are discussed.

4

ELECTRONIC WASTE: THE DARK SIDE OF THE HIGH-TECH REVOLUTION

> Future archaeologists will note that at the tail end of the 20th century, a new, noxious kind of clutter exploded across the landscape of the world—the digital detritus that has come to be called e-waste.
>
> Chris Carroll, "High-Tech Trash"

INTRODUCTION

This chapter addresses the health and environmental risks associated with electronic waste (e-waste) from cradle to grave. The issues of e-waste generation, handling, disposal, and export to underdeveloped nations for reuse, recycling, and disposal are discussed. The pattern of environmental injustice in the movement of e-waste especially from the global North to the global South of the world is addressed. First, the chapter presents selected definitions of e-waste, sources of e-waste, and some background literature concerning the growth and diffusion of electronic devices around the world in recent decades. The emergence of e-waste as a global environmental problem of immense proportions is introduced. Second, a conceptual understanding of what constitutes e-waste is provided, including a model of e-waste flow from the generators to processors and from environmental contamination to human health risks. Third, contrary to the image of electronic devices as environmentally benign, lean, and clean, the multitudes of toxic substances neatly packed inside these devices that pose a serious health threat to the public and the environment once discarded and dismantled are emphasized. The case of Guiyu, China, is presented to illustrate the adverse health and environmental impacts of e-waste. The need for stringent regulations to monitor safe handling of e-waste domestically and to control international flow of such waste is also discussed.

Defining E-Waste and Its Sources

Presently, there is no universally agreed-upon definition of e-waste. Different authors tend to define the term in their own way, especially by providing a catalogue of discarded electronic or electrical equipment. With this approach, e-waste encompasses a broad array of discarded or end-of-life (EOL) electronic devices ranging from personal computers (PCs), laptops, notebooks, televisions, cathode ray tube (CRT) monitors, cell phones, stereos, digital cameras, electronic games and their accessories, printers, scanners, copy machines, fax machines, geographical positioning system (GPS), personal data devices (PDDs), to major household appliances—such as microwave ovens, refrigerators, freezers, and security alarm systems. In short, e-waste refers to broken, obsolete, and undesirable electronic products destined for disposal or recycling. By function, e-waste consists of electronic equipment that have been used for data processing, telecommunications, word processing, or entertainment in households and businesses that have reached their EOL, and are now regarded as obsolete or outdated, broken, and not repairable. The Environmental Protection Agency (EPA) defines e-waste as electronic products that are approaching, or are at, the end of their useful life. Thus, e-waste generally implies any old, EOL electronic or electrical equipment that has been disposed of by its original owners because it is obsolete, broken beyond repair, or considered undesirable. This definition is preferred because of its broad scope of coverage of all conceivable EOL electronic/electrical waste types. Table 4.1 presents a summary of definitions of e-waste commonly found in the literature.

Three categories of e-waste can be designated: (a) large household appliances—such as dishwashers, refrigerators, washing machines, microwave ovens, etc., which are often referred to as "white goods"; (b) information technology and telecommunication equipment—encompassing desktop PCs, laptops, printers, fax machines, scanners, etc., known as "brown goods"; and (c) consumer entertainment/information equipment such as TVs, VCRs, MP3 players, telephones, cellular phones, stereo equipment, transistor radios, etc., referred to as "gray goods." Both (b) and (c) categories are comparatively more complex to recycle due to their toxic or hazardous constituents. While all the three categories are included in the definition of waste electrical electronic equipment (WEEE)/e-waste in the European Union (EU), only (b) and (c) categories are emphasized as e-waste in the United States. Nevertheless, these categories have certain components including metal, motor/compressor, cooling, plastic, insulation, glass, liquid crystal displays (LCDs), rubber, wiring/electrical, concrete, transformer,

Table 4.1   Common Definitions of E-Waste

Definition	Source
E-Waste is a term used to cover almost all types of electrical and electronic equipment that has entered or could enter the waste stream. It is also known as WEEE (i.e., waste electrical and electronic equipment).	STEP (2008)
E-Waste includes computers, entertainment electronics, mobile phones, and other items that have been discarded by their original owners because they were obsolete, broken, or beyond repair.	New World Encyclopedia
E-Waste refers to electronic products that are "near" or at the "end of their useful life."	US EPA
WEEE/E-Waste is defined as any appliance using an electronic power supply that has reached its end-of-life (EOL).	OECD (2001)
E-Waste encompasses a broad and growing range of electronic devices ranging from large household appliances such as refrigerators, air conditioners, cell phones, stereos, and consumer electronics to computers that have been discarded by their owners.	Basel Action Network (Puckett and Smith, 2002)
E-Waste is a term used loosely to refer to obsolete, broken, irreparable, or unwanted electronic equipment such as TVs, computers, and computer monitors, laptops, CPUs, printers, cell phones, copiers, fax machines, scanners, stereos, or video gaming systems, and associated wiring.	2009 CRS Luther Report for Congress

thermostat, brominated flame retardant (BFR)-containing plastic, batteries, CFC/HCFC/HFC/HC, external electric cables, refractory ceramic fibers, radioactive substances, lead, and electrolyte capacitors (Pinto, 2008: 66). Furthermore, many household appliances previously considered as electrical equipment, such as refrigerators, are now equipped with programmable microprocessors, thus elevating them to the status of electronic products. However, due to their longevity, large appliances are less common in the volumes of e-waste.

E-waste is generated by three key sectors in the United States, including individuals, households, and small businesses; big businesses, corporations, institutions, and government; and original equipment manufacturers (OEMs). How each of these sectors generates and manages e-waste is explained in the following sections.

Individuals, Households, and Small Businesses

Used electronic equipment including computers, stereos, TVs, and other audio and video consoles are frequently disposed of by individuals in households and small businesses because new technology has either rendered them obsolete, undesirable, or unfashionable; or because they are defective and have been out of warranty. As mentioned by Puckett et al. (2005: 6), with the useful life span of a PC currently at two years or less, coupled with the issue of incompatibility of new software with old hardware, customers are often forced to buy new ones and discard the old ones. Furthermore, with the exception of California and Massachusetts, most states still grant legal exemptions in the definitions of solid and hazardous wastes in which individuals, households, and small business owners can legally dump their computers, TVs, and other electronic devices into their garbage cans for disposal in the local municipal landfills.

Big Businesses, Corporations, Institutions, and Government

Big businesses, large corporations, public schools, colleges and universities, and government agencies upgrade employee computers on a frequent basis. Most of these entities with a large number of employees replace each computer about every two to three years. However, it is illegal for these large users to discard e-waste in landfills. Therefore, their obsolete computers and monitors, printers, copy machines, and other electronic devices are channeled to reuse, recycling, or export processing centers. Some of the used computers from institutions and government offices are often donated for reuse, or sold at public auctions in large quantities.

Original Equipment Manufacturers (OEMs)

The OEMs generate a huge volume of e-waste during the course of production (Grossman, 2006). For instance, whenever the units coming off the assembly line are defective or fail to meet quality standards, they must be discarded. Some OEMs handle their own e-waste management and disposal operations while others simply subcontract their e-waste management to recycling companies (Puckett et al., 2002). The fact that most OEMS in Silicon Valley, California, and other production sites have at least a National Priority List (NPL) site in their history attests to the fact that not all of them adhere strictly to proper management and safe disposal of their e-waste (see Grossman, 2006).

BACKGROUND

Over the past three decades, there has been an unprecedented explosion in the production, distribution, and use of electrical and electronic high-technology equipment around the globe (ITU, 2009a, b; UNEP, 2007). In the more developed countries (MDCs) of the global North, such as the United States, Canada, Western Europe, and Japan, electronic devices and electrical equipment, such as computers, printers, televisions, telephones, cell phones, fax machines, etc., have diffused among businesses, government, educational and religious institutions, and households. With vigorous advertisement, high-tech industry entices consumers with electronic devices and utilities that promise a more flexible, independent, and fun life. These devices now appear to be indispensable in the daily routines of people in an information society. Electronic equipment or high-technology products have become prominent features of life in postindustrial society (USEPA, 2007). In one way or another, electronic equipment is an integral part of everyday life—from televisions in homes, GPS and stereos in automobiles, portable devices including DVD players, MP3/4 players, iPads, iPods, iPhones, and cellular (mobile) phones in the possession of average individuals, to computers and laptops in businesses and educational facilities and most households, electronic devices are ubiquitous in modern society.

Table 4.2 shows the growing trends in electronic sales and use in the world since 1980. Table 4.3 presents the trends in the volumes of electronic products sold in the United States from 1975 to 2007. The sales of PCs and PC use per thousand have grown exponentially

Table 4.2 Global Growth in PC Sales and Use, Cell Phones, and Telephone Lines (Million Units)

Year	PC Sales	PC Use/1000	Cell Phone Subscribers/1000	Telephone Lines/1000
1980	0.76	–	–	–
1985	6.6	–	–	–
1990	9.5	18.7	2.1	99
1995	21.4	39.8	15.6	122
2000	46.0	86.1	123.0	161
2005	56.6	153.5	320.0	195
2010*	66.7	166.5	478.0	204

Projected Sources: Adapted from "Et Forecasts, World PC Market," http://www.etforecasts.com/products/ES.pcww1203.htm (accessed 12/30/09); Cellular Subscriber by Country, available at http://www.etforecats.com/products ES.cellular.htm (accessed 12/30/09).

Table 4.3 Volume of Electronic Products Sold in the United States, 1975–2007 (Millions)

Year	Total TVs	Total Desktops	Laptops	Cell Phones	Printers	CRTs, Monitors
1975	11.5	N/A	N/A	N/A	N/A	N/A
1980	16.8	1.5	N/A	N/A	1.8	1.2
1985	28.1	7.0	N/A	0.11	4.6	5.6
1990	23.2	14.4	N/A	2.6	9.6	10.3
1995	25.3	23.2	1.9	14.5	28.5	23.3
2000	37.4	40.9	6.6	71.2	46.4	37.5
2005	31.0	38.0	19.6	150.0	63.6	15.9
2007	28.6	34.2	30.0	181.9	N/A	–

Sources: Adapted from USEPA, "Electronics Waste Management in the United States: Approach 1," July, 2008, EPA 530-R-08-009; "Management of Electronic Waste in the United States: Approach 2," April 2007, EPA 530-R-07-0046.

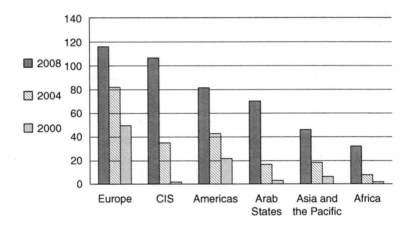

Figure 4.1 Mobile-Cellular Phone Penetration Rates.

Sources: International Telecommunication Union (ITU), Information Society Statistical Profiles

over the past three decades. In both tables 4.2 and 4.3, the cellular phone is the fastest-growing electronic device spreading around the globe like wildfire.

Figure 4.1 displays regional cell phone penetration rates from 2000 to 2008. Within this eight-year time span, virtually all regions of the world had adopted this technology at a very rapid rate. Even in Africa, cell phone subscription and use have accelerated at a phenomenal rate both in urban and rural areas. Although the rate of subscription varies

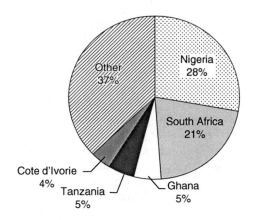

Figure 4.2 Distribution of Mobile-Cellular Subscriptions in Africa, 2008.

Sources: ITU, World Telecommunication, Information Society Statistical Profiles, 2009.

from country to country as depicted in figure 4.2., as of 2008, Nigeria has the highest subscription rate followed by South Africa, Ghana, and Tanzania. Incidentally, the volumes of e-waste seem to trail the distribution of cell phone penetration with Nigerian ports receiving the bulk of e-waste shipped to Africa. The market for both new and refurbished electronics in Nigeria has expanded significantly over the past decade (Schmidt, 2006; Umesi and Onyia, 2008; Puckett et al., 2005). As reported by the Basel Action Network (BAN), an environmental nongovernmental organization (NGO), an estimated 500 containers of used and junk electronic equipment arrive in Lagos every month; and each container has approximately 800 monitors and CPUs. The origins of the used electronic and e-waste received were discovered to be about 45 percent from the United States, 45 percent from Europe, and the remaining 10 percent from other countries such as Japan, South Korea, and Israel (Puckett et al., 2005: 12).

In the developing countries of the global South, there is an increasing demand for new, refurbished, used, and recyclable obsolete electronic devices that can be acquired at low prices. While China and India are rapidly closing the "digital divide" or technological gap between the developed and the developing countries, the growing accumulation of e-waste in both countries has raised serious health and environmental concerns (see Iles, 2004; Leung et al., 2006; Pinto, 2008).

Globally, the EOL aspects of electrical and electronic technologies have evolved into an accumulation of gigantic volumes of e-waste now characterized as e-waste tsunami, creating a cyber-age nightmare for many unsuspecting developing countries (Puckett et al., 2005).[1] Over 40 years ago, Gordon Moore, the cofounder of Intel, the computer chip manufacturer, observed that computer processing power doubles almost every two years. This "Moore's law," however, overlooks its corollary—that all the new powerful computers are at the same time on the verge of planned obsolescence. The useful life span of a computer has decreased from four to five years to only two years recently. According to UNEP (2007: 12), e-waste is one of the fastest-growing waste streams in the world today. About 20–50 million metric tons of e-waste are generated annually around the globe, accounting for more than 5 percent of all municipal waste.[2] The global e-waste market is forecast to reach 53 million tons by 2012.[3] About 60.1 percent of the total e-waste consists of metals including iron, aluminum, gold, silver, platinum, and toxic heavy metals such as arsenic, mercury, and lead.[4] In advanced nations of the global North, e-waste accounts for about 1 percent of the total solid waste on an average, and this is expected to increase to 2 percent by 2010 and beyond. In the United States alone, e-waste currently accounts for about 3 percent of the total volume of municipal waste generation. Nationally, an estimated 5–7 million tons of computers, televisions, stereos, cell phones, electronic appliances, and toys, as well as other electronic gadgets become obsolete and discarded annually. According to the EPA, e-waste is the fastest-growing component of the municipal waste stream in the United States with over 3 million tons disposed in 2007.[5]

The recent switch from analog TVs to digital high-definition televisions (HDTVs) in the United States has contributed to a significant spike in the volumes of HDTVs sold as well as old analog TV sets discarded or in storage for future disposal. Historically in the EU, e-waste or WEEE increases by 16–28 percent every five years, which translates into an increase that is three times faster than the average yearly municipal solid waste generation. E-waste generation within the EU is projected to grow at a rate of 3–5 percent per annum in the nearest future. According to Australia's Department of the Environment and Heritage's 2004–2005 report, approximately 9 million computers, 5 million printers, and 2 million scanners in businesses and households were replaced in 2004 with a substantial amount of these discarded as e-waste (Department of the Environment and Heritage, 2004–2005). E-waste in Australia is presently growing at more than three times the rate of municipal waste, with computers and TVs being the major components. About 37 million computers and 17 million TVs were either already disposed in landfills or sent to landfills in 2008.[6]

An upward trend in e-waste volume is also expected in several developing countries in Asia, Latin America, and Africa where undocumented trading in used electronics and e-waste is rapidly spreading. Illicit e-waste trading schemes and the proliferation of the underground economy in used electronics and electrical appliances in many countries such as China, Ghana, India, Nigeria, and Pakistan make collecting accurate data on volumes of e-waste traded extremely difficult. As mentioned by Puckett et al. (2005: 5), the data concerning trade in used electronics and e-waste are virtually nonexistent because of the fact that the harmonized tariff schedule (HTS) does not properly designate codes for e-waste other than batteries; thus, no one really knows the total volumes of e-waste traded globally. What has been reported, however, is the fact that about 75–85 percent of used electronic equipment—computer CPUs, monitors, CRTs, TVs, etc.—shipped to African countries such as Ghana and Nigeria is broken beyond repair or refurbishment. Lacking the technical capacity and infrastructure to process and manage these kinds of wastes, useless e-waste is dumped openly in heaps within residential neighborhoods where children play (see figures 4.3a and 4.3b). According to BAN coordinator Jim Puckett, who along with associates conducted

Figure 4.3a Unprotected E-Waste Dump Sites Posing Severe Threats to Children in Less Developed Countries (LDCs). Courtesy of the Basel Action Network.

Figure 4.3b   E-Waste Contaminated Dump Site Posing a Serious Danger to Children in Nigeria.

a field study of e-waste in Nigeria, enormous piles of e-waste abound throughout the countryside, mostly sent through Lagos. People were using e-waste to fill in swamps, and whenever the piles got too high, they would set them on fire (Puckett et al., 2005; Schmidt, 2006). Thus, the mishandling of growing e-waste volume in the country poses a serious threat to human health and the environment.

From a global political economy standpoint, e-waste serves as an example of the pattern of technology and material flows in the contemporary globalized economy, where capital and materials flow with little regard for geopolitical boundaries. Production and consumption systems increasingly withdraw raw materials, resources, and energy from the periphery to feed the core, and move pollution and adverse health effects around the globe. Grossman (2006: 5) notes that, each neatly packed electronic equipment has a story that originates in mines, refineries, factories, rivers, and aquifers, and ends on pallets, in dumpsters, landfills, and in poor people's backyards around the globe. Computers, TVs, and most other modern electronics and electrical equipment are designed in the United States, Western Europe, and Japan and manufactured in countries such as China and Taiwan, with raw materials and energy derived or extracted from Africa, Australia, Latin America, and the Middle East. The finished products are marketed and used all over the world, and upon reaching their EOL are sent either as real or surreptitious schemes of trading, recycling, and disposal to noncore nations such as China, Ghana, India, Nigeria, and other Third World countries (Iles, 2004). Clearly, the flow of e-waste is along the paths of least resistance around the globe. Most impoverished nations readily accept e-waste either to earn a little foreign exchange or in the hope of salvaging valuable constituents of these for recycling.

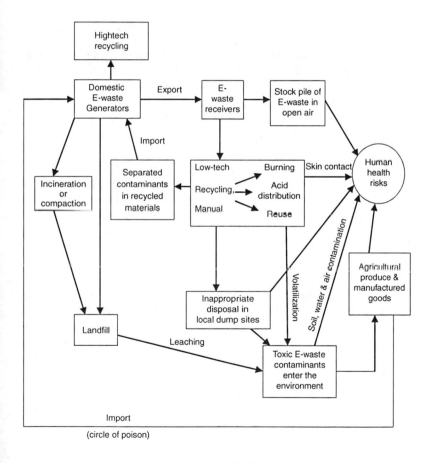

Figure 4.4 A Model of Global Movement of E-Waste from Generators to Receivers and Human Health Risks.

Figure 4.4 presents a model of the dynamics of global e-waste flow from domestic generators to disposal in landfills, incineration, or managed by high-tech recyclers to export to LDCs. Most LDCs with dismal or nonexisting e-waste management capacity often release e-waste contaminants into the environment, thereby contaminating agricultural produce and manufactured goods, such as jewelry and children's toys, and ultimately affecting human health and the health of other organisms. The import of e-waste-contaminated agricultural produce and manufactured goods by e-waste generators and others in MDCs completes the global circle of poison (see Weir and Schapiro, 1981).

The Magnitude of the E-Waste Problem

E-waste represents a rapidly growing national and transnational problem of immense proportions (Iles, 2004; Pellow, 2007; Silicon Valley Toxic Coalition [SVTC], 2009; GAO, 2008; Finlay, 2005). Using the phrase introduced by Kai Erikson, e-waste now constitutes "a new species of trouble" locally and across nations. According to the United Nations Environmental Program (UNEP) assessment, the global e-waste market is expected to grow from 7.2 billion dollars in 2004 to 11 billion dollars in 2009, with an annual growth rate of 8.8 percent. As mentioned earlier, globally, annual e-waste generation is projected to reach 53 million metric tons by 2012.[7] This phenomenal growth rate is due to the increasing planned "obsolescence rate" of most electrical and electronic equipment that propels their trading for material recovery or recycling and hazardousness (UNEP, 2007: 29). The recent technological revolution in electronics, computers, TVs, and the telecommunication industries producing more efficient, faster, and affordable products with more functionality and a greater appeal to the consumers has accelerated e-waste accumulation.

Due to economic and technological advances, it is often cheaper and more convenient to simply acquire a new electronic product than to upgrade or repair an old one. Recently, I took a small laptop purchased in the late 1990s to a computer shop in Austin, Texas, for repair and possible memory upgrade. The attending technician looked at it briefly and said "I think you're better off to go and buy a new one instead of wasting your money on a repair because you could almost buy a brand new one with the repair charges for the obsolete one." I listened to his advice and the laptop joined the piles of other household gadgets in our garage awaiting a new home.

Millions of homes across the United States accumulate such e-waste in their storage annually. It is estimated that consumers have an average of two to three obsolete computers, TVs, and other electronic gadgets in their garage, basement, closet, or storage place (UNEP, 2005). The consumers in postindustrial affluent society, most notably the United States and the EU, are generally reluctant to upgrade or purchase refurbished electronics.[8] People perceive buying new machines with warranty as cheaper than upgrading the existing ones. Iles (2004) contends that people have this perception simply because they are oblivious to the devastating effects of the production of computers and other electronic equipment on the environment and human health since the true costs of their production are externalized.

A recent study suggests that 1.3 kg of fossil fuels and chemicals are needed to manufacture a 2-gm memory chip. To produce a desktop computer and a 17" CRT monitor, at least 240 kg (530 pounds) of fossil fuels, 22 kg (48 pounds) of chemicals, and 1.5 tons (or 1,500 kg) of water are used (see UNEP, 2005; Kuehr et al., 2003: 64). Thus, it is indisputable that the manufacturing of computers and other electronic products has a huge ecological footprint in the biosphere. The natural resources withdrawn or extracted in the course of production of computers and other electronic goods are so intense that they are projected to have significant adverse effects on fossil fuel depletion as well as on global climate change (UNEP, 2005; Grossman, 2006; Pellow, 2007).

The accumulation of obsolete or unwanted electronic devices represents the inescapable outcome of the treadmill of production within the high-tech electronic industry and the insatiable appetite of the consuming public. Alan Schnaiberg and his associates offered the treadmill of production theory to explain why environmental degradation had increased phenomenally in the post–World War II period, especially in the United States. The key change emphasized in the theory was the preoccupation with accumulation of capital in Western economies through the replacement of labor with new technologies for the sake of profits. The new technologies required significantly more resources, energy, and hazardous chemicals to replace previous, more labor-intensive production processes, thus yielding deeper levels of ecological and social disorganization than ever before (Gould et al., 2008). The noxious and hazardous components of industrial metabolism are externalized and spread far and wide around the globe.

Just as electronic products have grown exponentially in the last decade or so, the amount of e-waste generated has also progressed in geometric proportion. As noted by Grossman (2006: 6), for the most part we have been so mesmerized by high tech, adopted its products with such alacrity, and been so busy thriving on its success and learning how to use the new PCs, laptops, PDAs, TVs, VCRs, DVD players, iPods, iPads, cell phones, etc., that, until recently, we have been unaware of the amount of waste and environmental degradation associated with these products. The United States generated over 3 million tons of e-waste in 2007, with the majority of the waste being shipped to underdeveloped nations. As previously mentioned, developing societies of the global South are the major destinations for e-waste from most of the global North countries including the EU, the United States, South Korea, and Japan. The disproportionate dumping of e-waste and other hazardous materials on the underdeveloped societies of the global South reflects a pattern of global environmental injustice. In addition to environmental contamination by the hazards and other xenobiotics present in the e-waste stream, public health and well-being are also put to risk.

Adverse Health Effects of E-Waste

From cradle to grave, electronic devices represent a complex mixture of materials, compounds, and components containing several hundreds of heterogeneous hazardous or toxic substances that pose a serious threat to human health upon exposure, and serious pollution to the environment when discarded in landfills or open dumpsites (UNEP, 2005). Among these are heavy metals such as antimony, arsenic, beryllium, cadmium, copper, chromium, lead, and mercury, some of which are extremely toxic.[9] The deleterious health effects of heavy metals have been discussed extensively in the previous chapter. Toxic heavy metals such as lead, cadmium, and mercury have been associated with neurological damage, kidney and liver dysfunction, skeletal problem, mental retardation, and developmental and reproductive disorders, among other serious adverse health effects. Flame retardants including halogenated compounds, such as polybrominated biphenyls (PBBs), polychlorinated biphenyls (PCBs), polybrominateddiphenyl ethers (PBDEs), and tetrabromo-bisphenol-A (TBBA), and other volatile organic chlorinated compounds, such as trichloroethylene (TCE) and tricholoroethane (TCA), employed in the manufacture of PCs and other electronic devices pose significant danger to human health and ecosystems when e-waste is disposed into landfills. These are among the POPs, that is, toxic chemicals that are lipophilic, hydrophobic, accumulate in human, animal, and fish tissues, and that travel around the globe. A number of these chemicals have been linked to thyroid dysfunction, endocrine disruption, mutagenicity, brain damage, and cancer (especially among experimental animals) (Schmidt, 2002; Thornton, 2000). Chapter 5 that follows is devoted to a more comprehensive discussion of these compounds and their effects on health and the environment.

In the United States, the high-tech computer-electronic industry has been linked to extensive contamination of the host communities and adverse human health conditions. For instance, Grossman (2006) identified a large number of leaking underground storage tanks (LUSTs) of toxic chemicals used by semiconductor plants in the Silicon Valley in California, which have extensively contaminated the land and underground water supplies in the area. Coincidentally, the county with the highest concentration of semiconductor and high-tech manufacturing plants also has the greatest concentration of Superfund sites than any other county in the United States.[10] More specifically, Grossman (2006: 79) contends that almost every large high-tech electronics and semiconductor manufacturer that began operations in the 1970s or earlier, has a Superfund site in its history. Furthermore, there are numerous sites not yet discovered or placed on the Superfund list across the country from Silicon Valley in California to Silicon Valley in Arizona, and to New Mexico, New York, Texas, and other states across the United States, where toxic contamination caused by chemicals and materials used in high-tech electronic manufacturing has led to persistent environmental and human health problems. Thus, untold levels of environmental health burdens are caused in the process of manufacturing high-tech electronic products.

The end of the life cycle of computers and other electronic equipment involves a series of steps including storage, reuse, recycling, and eventual disposal in landfills or open dump sites, as is commonly practiced in underdeveloped countries. Researchers have estimated that about 75 percent of all computers ever sold in the United States are stockpiled in the storage, waiting to be donated or sold for reuse and recycling, or to be completely disposed of (UNEP, 2005). At public auctions in the United States, pallets of mixed junk and obsolete, albeit reusable, computers from businesses, government establishments, colleges, and public schools are sold "as is" to the highest bidders, who turn around and sell or donate these as an act of charity to disadvantaged people in the United States or to people of the Third World. Often, these are channeled through domestic electronic recyclers who export used electronic goods directly to developing countries such as China, India, Pakistan, Ghana, and Nigeria.

NGOs such as Greenpeace, BAN, and the Silicon Valley Toxics Coalition (SVTC), and several scientists have devoted considerable attention to the growing hazards of e-waste proliferation in developing countries and the consequent risks to human health and the ecosystem (see Schmidt, 2006, 2002; Kahhat and Williams, 2009; Tsydenova and Bengtsson, 2009; Robinson, 2009; Puckett et al., 2005; Leung et al., 2006; Ma et al. 2009). As indicated by Widmer et al. (2005: 444), e-waste generally contains over 1,000 different constituents, many of which are extremely toxic. These represent potential threats to human health and the well-being of all other species in the environment. When e-waste scraps are improperly handled as found in many underdeveloped countries such as Ghana, Nigeria, Pakistan, India, and Peru, etc., their hazardous constituents are released into the air, soil, underground water, rivers, and streams. Resident populations are at greater risk of absorbing toxic chemical compounds through dermal contact, inhalation, and ingestion by drinking contaminated water and consuming agricultural produce from e-waste-polluted areas. Adverse health effects of these substances have been well documented in the literature (see table 4.4). Table 4.4 presents the inorganic toxic compounds and halogenated organic compounds found in the various components of electronic devices and e-waste. The recyclability of these elements and the potential acute and chronic health effects of each compound are presented—ranging from allergic reactions, breathing difficulties, to damage to vital organs, DNA, and cancer of various sites.

As investigated by Puckett et al. (2005), Brigden et al. (2005, 2008), Schmidt (2002, 2006), and Tsydenova and Bengtsson (2009), developing countries with lax or nonexisting environmental regulations and abundance of cheap labor are magnets for e-waste exported from developed countries for reuse and recycling. E-waste exporters from the United States and other developed nations basically circumvent the existing national environmental laws and the Basel Convention on transboundary movement of hazardous waste simply by labeling obsolete computers and electronics as products for

Table 4.4 Hazardous Substances in PCs, Recyclability, and Their Potential Adverse Health Effects

Substance	Use/Location	Recyclability (%)	Adverse Health Effects
Aluminum	Structural, conductivity/ housing, CRT, PWB*, connectors	80	Damage to kidney and central nervous system, skin rashes, skeletal problems, respiratory problems including asthma, linked to Alzheimer's disease
Antimony	Diodes/housing, PWB, CRT	0	Pneumoconiosis, heart problems, stomach ulcers
Arsenic	Doping agents in transistors/PWB, light-emitting diodes	0	Allergic reactions, nausea, vomiting, decreased red and white blood cell production, abnormal heart rhythm, (inorganic arsenic is a known human carcinogen)
Barium	Vacuum tube/CRT	0	Breathing difficulties, increased blood pressure, swelling of brain, damage to heart, liver, and kidneys
Beryllium	Thermal conductivity/ PWB, connectors	0	Lung damage, allergic reactions, chronic beryllium disease (beryllium is a suspected human carcinogen)
Cadmium	Rechargeable batteries, blue-green phosphor emitter/housing	0	Pulmonary and kidney damage, bone fragility, (cadmium is a suspected human carcinogen)
Chromium	Decorative, hardener/ (steel) housing	0	Ulcers, convulsions, liver and kidney damage, asthmatic bronchitis, DNA damage, carcinogenic
Cobalt	Structural, magnetivity (steel) housing, CRT, PWB	85	Lung and heart effects, dermatitis, liver and kidney problems
Copper	Conductivity/CRT, PWB, connectors	90	Chronic exposure can irritate nose, mouth, and eyes and cause headaches, dizziness, nausea, and diarrhea
Europium	Phosphor activator/ PWB	0	Cancer of the liver and bone
Gallium	Semiconductors/PWB	0	Carcinogen in experimental animals
Germanium	Semiconductors/PWB	0	Carcinogen in experimental animals
Gold	Connectivity, conductivity, PWB, connectors	99	
Indium	Transistor, rectifiers/ PWB	60	Damage to the heart, kidney, liver, and may be teratogenic

Continued

Table 4.4 Continued

Substance	Use/Location	Recyclability (%)	Adverse Health Effects
Lead	Metal joining, radiation shield, CRT, PWB	5	Damage to central and peripheral nervous system, kidneys, and brain development
Manganese	Structural, magnetivity (steel), housing	0	
Mercury	Batteries, switches/housing, PWB	0	Chronic brain, kidney, lung, and fetal damage (see chapter 3)
Nickel	Structural, magnetivity (steel), housing, CRT, PWB	80	Allergic reactions, asthma, impaired lung function, chronic bronchitis, carcinogenic
Niobium	Welding alloy/housing	0	May cause kidney damage and scarring of the lung if inhaled
Palladium	Connectivity, conductivity/PWB, connectors	95	Damage to bone marrow, liver, and kidneys; may also cause skin, eye, and respiratory tract infections
Platinum	Thick film conductor/PWB	95	Cancer, damage to kidneys, intestines, hearing, and bone marrow
Rhodium	Thick film conductor/PWB	50	Carcinogen, central nervous system dysfunction in animals
Ruthenium	Resistive circuit	80	Yet to be determined, may cause skin and eye irritation
Selenium	Rectifiers/PWB	70	
Silver	Conductivity/PWB, connectors	98	Breathing problems, skin, lung and throat irritation, stomach pain
Tantalum	Capacitors/PWB, power supply	0	Eye and skin irritation
Terbium	Green phosphor activator, dopant/CRT, PWB	0	Eye and skin irritation, other effects are yet to be investigated
Tin			
Titanium	Pigment, alloying agent/housing	0	No conclusive evidence
Vanadium	Red phosphor emitter/CRT	0	Damage to lungs, throat and eyes, possible kidney and liver dysfunction
Yttrium	Red phosphor emitter/CRT	0	Damage to lungs and liver of animals reported
Zinc	Battery, phosphor emitter/PWB, CRT	60	Very high levels can damage the pancreas, and it is dangerous for unborn and newborn children

* *Note*: Printed Wiring Board.

Sources: Information compiled from the ToxFAQs, Agency for Toxic Substances and Disease Registry (ATSDR), Atlanta, GA. Available online at http://www.atsdr.cdc.gov/toxprofiles/phs54.html (accessed 12/20/09); Microelectronics and Computer Technology Corporation (MCTC), *Electronics Industry Environmental Roadmap* (Austin, TX: MCTC, 1996).

Figure 4.5 A Nigerian Repairman Dismantling a CPU and Monitor Attempting Repairs from Dismantled Parts While Generating E-Waste from Unusable Parts.

Source: Courtesy of the Basel Action Network.

reuse and recycling. Comingling of used electronics with scraps is another technique used quite often to evade regulation. In a recent report by the US Government Accountability Office (GAO), existing US hazardous waste regulations have not deterred exports of potentially harmful electronics, primarily because the existing EPA regulations on e-waste are narrowly focused (only on CRTs) and companies generally circumvent the CRT rule, and also because of lack of EPA enforcement (GAO, 2008).

According to the BAN and GAO reports, Nigeria has no capacity for materials-recovery operations for copper, lead, steel, precious metals, plastics, etc., or collection mechanisms for e-waste. Consequently, most imported obsolete or junk electronics and computers are mainly for reuse, and those that are perceived to be useless are disposed of in local dumps that are not sanitary (GAO, 2008; Puckett et al., 2005). The piles of junk computers, electronic scraps, and plastic encasements are often burned openly along with other types of waste. In both Nigeria and Ghana, e-waste such as TVs, computer CPUs, CRTs, etc., are manually disassembled in an open space or under a shed by local repairmen using rudimentary tools such as sledgehammers and screwdrivers as shown in figure 4.5. Open burning of plastic encasements of TVs and computers, plastic wires and cables, and the constituent BFRs produces toxic by-products of dioxins and furans as mentioned earlier. E-waste handlers in underdeveloped countries including Ghana and Nigeria generally operate without wearing any protective gear and consequently risk bodily contamination through dermal contact, inhalation, or ingestion of toxic elements from e-waste.

Most African countries lack awareness of the dangers posed by e-waste and e-waste collection and disposal systems. Hence, discarded e-waste are treated in the same way as ordinary household solid waste and discarded in neighborhood dump sites, all of which are unlined, unmonitored, and close to groundwater. Since the e-waste contains toxic elements, the human health and environmental risks posed by these upon degradation or when openly incinerated are of major concern. As mentioned in chapter 3, lead and mercury are highly potent neurotoxins, especially among children, who may experience mental retardation or IQ deficits, or developmental and behavioral problems even at low levels of exposure (see Schmidt, 2002: A192). Leung et al. (2008) recently note that exposure to high levels of heavy metals may cause acute and chronic toxicity, including damage to the central and peripheral nervous

systems, blood composition, kidneys, liver, and lungs, and may even
result in death.

THE CASE OF GUIYU, CHINA

Guiyu is a town of about 150,000 residents in Guangdong Province
of Southeast China. It is now regarded as the e-waste recycling capital of the world (Robinson, 2009). Since the early 1990s, this town
and the surrounding villages have been changed from a poor, rural,
rice-growing area to an e-waste boom town. The bulk of e-waste
exported from the United States, the EU, and Japan typically ends up
in this town, recently transformed from a peasant agricultural subsis-
tence to an e-waste recycling enclave. In addition to imports, China
generates over 1.7 million tons of e-waste annually, which includes
approximately 5 million TV sets, 4 million refrigerators, 5 million
washing and drying machines, 10 million mobile phones, and 5 million PCs, most of which are sent directly to Guiyu for recycling or
dumping (Bodeen, 2007). The volume of e-waste generated in China
is expected to increase to 5.4 million tons by 2015.[11]

A significant number of previous Guiyu farmers now eke out their
living as e-waste laborers working as disassemblers, using crude tools
and methods to recover precious metals and other valuable components. About 80 percent of families in the town are involved in
e-waste recycling facilities, and approximately 10,000 migrant workers are employed in the industry (Yu et al., 2006). This town has been
studied by several scholars and NGOs particularly interested in the
environmental and health impacts of e-waste recycling activities (see
Leung et al., 2006, 2008; Brigden et al., 2005; Puckett et al., 2002;
Huo et al., 2007; Robinson, 2009, Yu et al, 2006).

Even though rice is still cultivated by a declining number of
farmers in the area, most of the available building space in Guiyu
has been converted to e-waste recycling structures (Puckett et al.,
2002). In a *New York Times* article, Bodeen (2007: 1) describes
Guiyu as an e-waste heartland with the following observations:
"The air smells acrid from squat gas burners that sit outside homes,
melting wires to recover copper and cooking computer mother-
boards to release gold. Migrant workers in filthy clothing smash
picture tubes by hand to recover glass and electronic parts, releasing as much as three kilograms of lead dust." As observed by Yu
et al. (2006), in Guiyu, piles of wires and plastics are burnt in the
open in proximity to rice fields or riverbeds to recover valuable

metals, and circuit boards are melted over coal grills to release chips; the undesirable residuals are then disposed in the rivers, rice fields, ponds, irrigation canals, and open fields, thus polluting the environment.

As noted by Leung et al. (2008: 2675), the extraction of electrical components and solder recovery from circuit boards are carried out mainly in family-run shops in residential areas. In fact, circuit board components litter the roads in Beilin district and other parts of Guiyu. Consistent with Bodeen's (2007) observation, Leung et al. (2008) also note that the entire area is characterized by distinguishable acrid fumes and malodorous discharges from molten solder, which irritate the eyes and the throat. Undoubtedly, the e-waste recycling activities in Guiyu have caused severe environmental pollution and serious adverse health conditions among the residents. Recent empirical studies of the town and its vicinities have revealed that the air, soil, water, and river sediments in the area are severely contaminated by toxic compounds released from e-waste recycling operations (Wong et al., 2007; Yu et al., 2006; Leung et al., 2006).

The residents of Guiyu have been found with higher incidence of e-waste-induced morbidity—including skin disorders, headaches, nausea, vertigo, chronic gastritis, and gastric and duodenal ulcers relative to other towns in southeast China not involved in e-waste recycling activities (Huo et al., 2007: 1113). Children are particularly at greater risk of exposure to toxic compounds at e-waste recycling centers in Guiyu. In a study conducted by Huo and associates, analysis of blood samples from 165 children aged one to six years in Guiyu found blood lead levels ranging from 4.40 to 32.67 µg/dL with an average of 15.3 µg/dL, which was significantly higher than that of their counterparts in towns without e-waste recycling operations. Children living in a community hosting e-waste workshops specializing in equipment dismantling, circuit board baking, and acid baths exhibited the highest blood lead levels.

In a preliminary study conducted by Brigden and associates, single dust samples obtained from the homes of two e-waste recycling workers in Beilin, China, exhibited higher levels of copper, lead, tin, antimony, and cadmium in the range of 4–23 times greater than levels found in neighboring homes without e-waste recycling workers.[12] The evidence reveals that e-waste recycling workers in Guiyu and the surrounding areas inadvertently carry toxic contaminants from e-waste workshops to their homes, thus increasing the exposure of children and all other family members to heavy metals and other e-waste xenobiotics. The US Centers for Disease Control and Prevention set elevated blood lead levels at $\geq$ 10 µg/dL in children aged six years or less (CDC, 1991). In children under six years of age, lead poisoning is a serious concern as it may cause lifelong learning disabilities, hearing impairment, speech problems, IQ deficits or mental retardation, behavioral problems including attention deficit disorder, hyperactivity, and aggressive or violent behavior, and in more critical cases comas, convulsions, and even death. The e-waste processing procedures in Guiyu that pose serious risk to the environment and human health are described next.

CRUDE E-WASTE PROCESSING PROCEDURES IN GUIYU

The formal mechanism for collection, processing, and disposal of e-waste is still rudimentary and in its infancy in China. Thus, e-waste collection, recycling, and disposal are carried out by small, informal, and unlicensed operators. In many parts of Guangdong province, including Guiyu, the large majority of e-waste operations are in people's backyards or small workshops. Based on close observation, Huo et al. (2007: 1114) describe the crude or primitive e-waste processing procedures in Guiyu as involving a series of steps, which include the following:

- Obsolete electronic equipment is dismantled using electric drills, sledgehammers, screwdrivers, and cutters to separate the constituent parts—that is, monitor, CD drive, hard drive, circuit boards, wires, cables, battery, transformers, chargers, plastics, and metal frames and casings that are sold for reuse or to other recycling shops for further processing.
- Circuit boards of computers and other electronic appliances are cooked over coal grills, or gas or kerosene burners, to melt the solder to release valuable components such as diodes, resistors, and microchips.
- For cellular phones and other portable hand-held devices, their circuit boards are separated using an electrothermal machine, which represents a source of human health and environmental contamination.
- The use of acid baths is prevalent in Guiyu. Some microchips and computer parts are typically soaked in acid baths to extract precious metals such as gold, palladium, silver, etc., and the residual acids are

disposed of in open fields, ponds, streams, and river banks, thereby contaminating the environment and elevating the risk of acid poisoning among the resident population.

• Plastics from e-waste such as polyvinyl chloride (PVC), acrylonitrile butadiene styrene copolymer (ABSC), and high-density polyethylene (HDPE) are sorted by laborers based on their rigidity, color, and luster. Plastic scraps not recyclable are put in a pile to be burnt. As mentioned earlier, dioxin and furans are released into the environment as the by-products of plastic and BFR burning. Those plastics considered acceptable for reprocessing are fed into grinders that release tiny pieces of plastic.

• To sort, extract, and reprocess valuable metals, transformers, chargers, CRTs, and batteries are forced open with hammers.

While e-waste processing procedures in Guiyu, China, are far more sophisticated than those in other LDCs, the techniques used are still considered crude or primitive at best, releasing toxic materials into the environment. Polybrominateddibenzo-p-dioxins and dibenzofurans as the by-products of e-waste incineration or open-burning in Guiyu and elsewhere have received considerable attention (Ma et al., 2009).

Controlling Global E-Waste Flow

How to control the global flow of e-waste, especially from affluent nations to underdeveloped countries, represents a serious challenge to the international community. Unlike other types of hazardous wastes, e-waste has a substantial amount of precious metals and other valuable constituents for recycling. Thus, with regard to e-waste, someone's trash is often another person's treasure. As shown previously in table 4.4, the recyclability rate for copper, cobalt, gold, platinum, and silver is 90, 85, 99, 95, and 98 percent, respectively. While acknowledging the adverse health and environmental effects of e-waste, a number of scholars have indicated the need to also focus attention on the functions of e-waste and the reverse supply chain, especially for PCs. According to Williams and collaborators, a reverse supply chain connotes the network of activities directly connected with the reuse, recycling, and disposal of products and their components.[13] The e-waste industry and other aspects of the reverse supply chain serve the following useful purposes: (a) generation of employment opportunities in the refurbishment and recycling industries; (b) bridging or mitigating the digital divide between more affluent and less affluent societies by increasing people's accessibility to affordable used PCs and other electronics; (c) extending the useful life of PCs and other electronics with the consequent relief on ecological withdrawal for manufacturing new equipment; and (d) in many developing nations, e-waste recycling and reverse supply chain represent a major source of income for poor communities.[14]

As observed by Iles (2004), numerous reverse-chain entrepreneurs have developed in several Asian countries capitalizing on profitable opportunities in e-waste recycling and other related activities. A vigorous demand for used or refurbished electronics and recycled materials in several industrializing nations in Asia seems to be a major magnet drawing e-waste to the region. Illes (2004: 82) further asserts that some governments in developing countries are willing to accept e-waste as a way of acquiring needed raw materials (as well as refurbished or used computers) to bridge the digital divide. Consequently, the demand for working used PCs, laptops, and CRTs in developing countries is growing, making a complete control of the movement of these products from the global North to the global South quite difficult. Several NGOs have monitored the global flow of e-waste around the globe, noting that e-waste flows along the routes of least or no resistance from MDCs to LDCs. With the exception of Australia, all the countries initiating reverse supply of used PCs or e-waste are in the global North—including the United States, the EU, Japan, and South Korea. Known destinations for e-waste dumping include Brazil, China, India, Mexico, Nigeria, Pakistan, Singapore, and Thailand.

In the United States, products with CRTs such as TVs and computer monitors that contain significant volume of lead and copper and other toxic materials are regulated as hazardous waste and their exports are controlled by the EPA.[15] As of July 2006, the EPA's CRT rule requires the exporter to notify the agency of intent to export such material for reuse or repair. Furthermore, in the case of CRTs exported for recycling, exporters are required to obtain consent from the importing country for shipment, consistent with the Basel Convention. E-waste and used electronic devices other than CRTs are not considered as hazardous waste under the RCRA of 1976 as amended, which is the major legislation governing hazardous waste handling and disposal in the country. According to a recent report by the US GAO, US hazardous waste regulations have failed to prevent exports of potentially hazardous used electronics due to lack of stringent EPA enforcement among other reasons. Hazardous waste regulations do not consider most used electronic devices—including computers, printers, cell phones, and copiers as hazardous, despite the fact that they can be mismanaged in the underdeveloped-nation destinations and can cause precarious health and environmental harm. Also, the GAO report underscores how US regulatory control measures are easily circumvented leading to unlawful exports of broken, nonworking CRTs. Even the EPA acknowledges the fact that a vast majority of used electronics including their component parts, donated for reuse or recycling are exported, both responsibly and irresponsibly (see GAO, 2008: 9).

Among other attempts to control international movements of hazardous wastes are the Basel Convention and the EU Directive on WEEE and the Restriction of Hazardous Substances (RoHS). The Basel Convention on the Control of Trans-boundary Movements of Hazardous Wastes and Their Disposal, was negotiated under the aegis of the UNEP between 1987 and 1989.[16] Several underdeveloped countries who are parties to the convention as well as NGOs such as Greenpeace advocate a global North-global South total ban on hazardous waste transfers while most of the Organization of Economic Cooperation and Development (OECD) nations prefer a regulatory system based on notification and consent. Regulation over a complete ban of hazardous waste was chosen by March 1989 when it was signed. The United States remains the only advanced industrialized country yet to sign or ratify the convention. As of 2010, the convention has 172 parties from all around the globe.[17]

The convention represents the most comprehensive international environmental agreement on hazardous and other types of waste. Among the primary objectives of the convention are to minimize the generation of hazardous wastes and to regulate and reduce their transboundary transfers with the ultimate aim of protecting human health and the environment against the adverse impacts from the generation, handling, management, transnational shipments, and disposal of such wastes.[18] Under the Basel Convention regime, hazardous waste exports are permitted only if prior notification and consent requirements are met by the exporting parties. Member parties are required not to export hazardous wastes to another party unless they can demonstrate that a competent authority in the receiving country has been duly notified and has consented to the import in advance of any waste shipment (Krueger, 2001, Kummer, 1998). Furthermore, notwithstanding the consent of the proposed country of import, the Basel Convention requires that nations of export prohibit transboundary shipment of hazardous wastes if there is any reason to believe that the wastes will not be handled in an environmentally sound manner in the receiving country. The convention further articulates an obligation of the nations of export to ensure that transnational shipments of wastes are accepted for reimport if those shipments fail to conform to the terms of exports (see Wirth, 1998: 238).

The EU WEEE and RoHS in electrical and electronic equipment directives represent the most progressive action till date in e-waste regulation. These directives governing WEEE and their contents were adopted by the EU and became effective in February 2003. The legislation provides for the creation of collection schemes where consumers can return their used electronic equipment or e-waste free of charge; the key objective here is to increase the recycling and reuse of electronic and electrical products. Also, the aim of these schemes is to substantially reduce the amount of e-waste entering incinerators and landfills and to eliminate hazardous substances in electronic equipment. For instance, RoHS requires heavy metals such as lead, mercury, cadmium, and hexavalent chromium and flame retardants such as PBBs or polybrominateddiphenyl ethers (PBDEs) to be substituted by safer alternatives.[19] As indicated by the European Commission, despite the directives on collection and recycling, more than two-thirds of e-waste is still being sent to landfills and to substandard waste-treatment facilities in or outside the EU. Illegal trade of WEEE to non-EU countries in Africa and Asia continues unabated. As Pellow (2007: 218) points out, a perverse effect of the WEEE directive is that as more material is collected for recycling, it may stimulate a greater demand to export e-waste illegally to underdeveloped nations for dirty recycling. Even though EU nations have signed the Basel ban on toxic waste exports, serious doubts remain about its enforcement as e-waste and other hazardous wastes continue to flow from the United States, the EU states, and other industrialized nations of the global North to developing nations of the global South, raising concerns about the global digital environmental and social injustice.

Future Directions in E-Waste Generation

Using available social indicators, it is projected that global e-waste production will increase as economies grow and new technologies are produced. A linear correlation has been found between the number of computers, other e-waste items, and per capita gross domestic product (GDP). Thus, economic prosperity is inextricably linked to volumes of PCs and other electronic equipment purchased, and the amount of e-waste generated. In some parts of the globe, e-waste is growing at a faster rate than the GDP. As Robinson (2009: 185) suggests, if the number of computers acquired is an indicator of total e-waste productions, then one can expect several industrializing nations in Eastern Europe, Latin America, and Asia to become major e-waste generators in the next 10 years. China is already a major producer of e-waste as mentioned before. The extent to which the existing regulations can curb the production and global flow of e-waste in the nearest future remains uncertain.

CHAPTER SUMMARY

This chapter has examined various dark sides of high-tech e-waste from cradle to grave both within the United States and across the globe. The growing evidence of ecological degradation associated with e-waste generation, mismanagement, and improper disposal along the paths of least resistance domestically and internationally has been discussed. The flow of e-waste from the affluent global North to the less developed global South continues unabated. The environment, wildlife, and health and well-being impacts of the multitudes of toxic materials from e-waste dumping are of increasing concern around the globe. Developing countries with rudimentary e-waste recycling infrastructure and those lacking such infrastructure but receiving huge volumes of e-waste are at greater risks of toxic contamination. The potential health risks due to e-waste xenobiotics have been addressed in this chapter. A case study of Guiyu was presented to illustrate the adverse consequences of e-waste toxic releases to human health and the biophysical environment. For the most part, e-waste and other reverse-chain products represent "opportunity" for some and a serious threat to others—especially those who may not be directly involved in the industry. The potential adverse impacts of hazardous by-products of e-waste on the environment, wildlife, and human population are of growing concern.

Global policy and legislative responses to the issue of transboundary movements of hazardous wastes have also been addressed in this chapter. The Basel Convention on the Control of Trans-boundary Movements of Hazardous Wastes and Their Disposal was examined along with the EU Directives on e-waste and hazardous waste. These represent noble efforts, albeit not necessarily sufficient in curbing the flow of waste from the global North to the global South. How economic growth and lack of adequate enforcement of existing regulations are driving the proliferation of illicit trade in used electronics and e-waste across the globe, have been addressed. Obviously, stronger legislative responses, enforcement of existing instruments, and cooperation among parties to the Basel Convention would be required to mitigate illegal flow of hazardous materials around the world.

5

Environmental Health Risks of Persistent Organic Compounds

Everyday life is "blind" in relation to hazards which threaten life and thus depends in its inner decisions on experts and counter-experts. Not only the potential harm but this "expropriation of the senses" by global risks makes life insecure.

Ulrich Beck

Introduction

The environment, defined as all the biotic and abiotic factors surrounding a given population, has a significant influence on the health and well-being of the population. In fact, the principal factor driving the concerns about environmental quality is its connection to human health. Directly through exposure to xenobiotics and indirectly through systemic environmental events, many health problems are inextricably connected to environmental factors (Smith, 2001). Different aspects of the environment—including the biological (biotic), physical (abiotic), social, cultural, and technological factors—affect the health status of the human population as well as other species within the ecosystems. Evidence in the literature suggests that environmental pollution and ecological degradation have a tremendous negative impact on people's well-being (Yassi et al., 2001). Polluted environments increase the probability or risk of exposure to contaminants, disease vectors, and other agents that may induce illnesses both for human and nonhuman species. Persistent organic pollutants (POPs) represent a significant threat to the environment and the health of all organisms including humans. Given the pervasiveness of risks in every sphere of social life, it is now widely recognized that we are living in a society of self-endangerment, self-injury, and potential self-annihilation (Strydom, 2002).

The purpose of this chapter is to address the health and environmental problems related to synthetic toxic chemical compounds, especially the key POPs of increasing concern at the local, national, and global levels. The fundamental properties of these chemicals and the established and potential health problems associated with each compound are discussed. The application of the precautionary principle by the international community to restrict or completely ban these chemicals, or place stringent regulatory measures on them, will also be discussed in the latter part of the chapter. Following the introduction, the background literature about the paradoxical benefits and risks of science and technology over the past century is discussed. Next, the basic properties of POPs are reviewed. The subsequent section is devoted to the pathways of exposure and health problems associated with POPs. The last section focuses on the precautionary principle as a policy tool and the Stockholm Convention on POPs, followed by a summary and the concluding remarks.

THE PARADOX OF TECHNOLOGICAL PROWESS, HEALTH, AND ENVIRONMENTAL RISKS

Even though environmental hazards and risks have always been present in societies all over the world, the technological and chemical revolutions of the nineteenth and the twentieth centuries have increased the levels of toxic materials and associated health problems to an unprecedented level in human history (Beck, 1992; Epstein et al., 1982; Thornton, 2000; Graham and Miller, 2001). Scientific and technological breakthroughs in the synthesis, production, and release of heterogeneous toxic chemical compounds have contributed to remarkable prosperity on the one hand, and on the other hand new arrays of unexpected dreadful health problems have been introduced by these chemicals. As indicated by Hofrichter (2000), exposure to an expanding array of toxic compounds in the air, water, and soil at places where people live, work, and play in communities, poses an increasing short-term and long-term threat to public health. The case of POPs such as dichlorodiphenyltrichloroethane (DDT), in particular, is paradoxical; while millions of lives have been saved through its application to control malaria, several millions have also suffered serious adverse health effects from acute and chronic exposures and toxicity (see Epstein et al., 1982; Carson, 1962; Wargo, 1996; Jones and deVoogt, 1999; Crinnion, 2000; Thornton, 2000; McGinn, 2002).

In our anthropocentric quests to conquer and subdue nature, arrest infectious agents, exterminate unwanted/undesirable species, and extend the carrying capacity of the Earth, synthetic organic chemical compounds were introduced shortly after World War II.[1] For more than half a century, a wide array of synthetic organic chemical compounds has been released into the environment. Most of these chemicals were initially greeted with enthusiasm and praised as "modern miracles" or as a boon, due to their effectiveness in controlling pests, improving agricultural yields, increasing the aesthetics of lawns and gardens, and their versatility in various industrial applications. It was not long until Rachel Carson (1962) sounded the alarm about the toxic and persistent nature of these chemicals and their adverse health effects on the ecosystem. Carson (1962: 8) poignantly emphasized:

> The central problem of our age has become the contamination of our total environment with such substances of incredible potential for harm—substances that accumulate in the tissues of plants and animals and even penetrate the germ cells to shatter or alter the very material of heredity upon which the shape of the future depends.

Thus, as indicated by Crinnion (2000), the twentieth century with its promise of prosperity and better quality of life through the application of science and technology, also brought a host of toxic chemical-related disasters, illnesses, and associated human sufferings. Kai Erikson (1994) summed these up in his book entitled *A New Species of Trouble*. Hofrichter (2000) contends that we now live in a toxic culture—a culture that degrades human and environmental health as the life world is colonized and exploited by the social systems. Toxic culture emerges with the development of social arrangements that encourage, facilitate, and excuse or rationalize the deterioration of the environment and human health as an inescapable aspect of economic development.

As noted by Myers (2002: 4), scientific knowledge of the impacts of toxic chemicals on health and the environment lags behind our knowledge and ability to synthesize these chemicals. Furthermore, traditional risk assessment is limited because it permits commercialization, distribution, and use of these products without a complete understanding of their adverse effects. Among the consequential problems are pervasive environmental contamination and diminution of human health after exposure. Epidemiology as a tool for developing protective standards is also limited by the fact that it is only applicable ex post facto, that is, after an epidemic has already occurred. Also, it is strongly biased toward negative results, contrary to what popular epidemiology or the lay public might suggest.

A growing number of scientists, international organizations, and nongovernmental organizations (NGOs) have devoted a considerable amount of time, effort, and energy in addressing the hazards and risks posed by persistent organic compounds (see Colborn et al., 1996; Jones and deVoogt, 1999; Wania and MacKay, 1996, 1999; Baskin et al., 2001; Thornton, 2000; Lallas, 2001, 2002; Schafer et al., 2001; Yassi et al., 2001; WWF, 1999; The World Bank and CIDA, 2001). For the past three decades, health problems associated with toxic chemical releases into the environment have been a growing major concern in societies across the globe. Specific cases of toxic chemical contamination that have raised people's level of concern range from the episodes at Love Canal, New York, Woburn, Massachusetts, and the accumulation of toxic substances in the Great Lakes and in the Cancer Corridor of Louisiana, to pesticide poisonings in Costa Rica, the accidental releases of dioxin in Seveso, Italy, in 1976; and from a similar contamination in Times Beach, Missouri, in the early 1980s, to the toxic chemical disaster at the Union Carbide pesticide plant in Bhopal, India, in 1986, which killed more than 2,000 and injured over 100,000 people, and the toxic chemical contamination in Koko, Nigeria, in the late 1980s (some of these cases are discussed in more detail in chapter 6; see Levin, 1982; Epstein, 1982; Wargo, 1996; Brown and Mikkelsen, 1990; Adeola, 1996, 2000).

In a foreword to *Our Stolen Future*, the former vice president of the United States, Al Gore, indicates that Colborn et al. (1996) build upon the work of Carson by reviewing a comprehensive and growing body of scientific evidence linking synthetic organic chemicals to a wide array of terrible health problems. Specifically, the endocrine-disruption hypothesis and the reviewed evidence that synthetic chemicals were acting like the hormone estrogen and causing reproductive and behavioral pathologies in humans and wildlife were publicized (see Baskin et al., 2001; Colburn et al., 1996; Wargo, 1996). The health effects of these chemicals are discussed in greater detail in the latter part of this chapter. At this juncture, it is essential to gain some understanding of the fundamental properties and characteristics of POPs.

BASIC PROPERTIES OF POPs

Due to more than half a century of extensive production, use, and release, POPs are now ubiquitous in the air, soil, and water. The major sources of air pollution contributing to the accumulation of POPs include the manufacture and use of certain pesticides, the production and use of certain toxic chemicals, and the unintentional formation of certain by-products of incineration, combustion, metal production, and mobile sources (Ballschmiter et al., 2002: 274). There is hardly any biome and species on the Earth left untouched by these chemicals. Scientific evidence suggests that all living organisms on the Earth presently carry measurable levels of POPs and related chemicals in their bodies. For instance, POPs have been found in marine mammals at levels concentrated enough to classify their bodies as hazardous waste. Scientists have reported evidence of POP contamination in our food, in human blood, and in breast milk (see Solomon and Schettler, 1999; Colburn et al., 1996; Thornton, 2000; Wargo, 1996; Wania and Mackay, 1996; Schafer et al., 2001).

By nature, POPs are organochlorine compounds with extensive longevity in the environment. They represent one of the most harmful classes of pollutants manufactured and released into the environment by humans, and as such, they are of particular relevance to human health and the health of other organisms in the environment (Moser and McLachlan, 2001). As noted by Eckley (2001: 26), POPs are characterized by their persistence in the environment with a tendency to bioaccumulate in the food chain, and their capacity for a long-range, transboundary dispersion poses a great threat to human health and the environment globally. A list of the 12 most dangerous POPs is presented in table 5.1 including aldrin, chlordane, DDT, dieldrin, endrin, heptachlor, hexachlorobenzene (HCB), mirex, toxaphene, polychlorinated biphenyls (PCBs), dioxins, and furans, collectively referred to as the "dirty dozen."[2] Their uses, longevity (half-life), and the known health effects are also displayed in the table.

The specific characteristics of POPs that warrant increased concern and the need to take immediate precautionary measures, to ban or restrict further production and use of these chemicals by the world community, include their extensive half-life (persistence), toxicity, lipophilic (fat-soluble) property, and bioaccumulative properties, as well as their ability to travel across the globe. These properties are elaborated in the following sections.

Toxicity and Longevity

POPs are extremely toxic chemicals with acute and chronic effects on pests, wildlife, and humans upon exposure. In fact, toxicity was their original virtue. Partly due to their toxicity, these chemicals resist breakdown by the natural processes and as such, remain within the environment for a long duration. As shown in table 5.1, most POPs persist in the environment for up to 23 years or more (some may take as long as a century to breakdown completely). For instance, chemical compounds such as DDT, endrin, HCB, mirex, polychlorinated-dibenzo-p-dioxins, and furans (a product of incomplete combustion) remain toxic and active for approximately 10–23 years in the soil, in fatty tissue, and in other environmental medium (see Wania and Mackay, 1996; Epstein et al., 1982; Eckley, 2001; Jones and deVoogt, 1999). Even though the 12 major POPs have now been banned or restricted in most industrialized countries, these chemicals continue to be produced and exported to Third World countries where regulations are lax. However, the "circle of poison" thesis suggests that "what goes around comes around," and POPs in particular, do not respect geographical boundaries (Weir and Schapiro, 1991).

Lipophilic and Bioaccumulative Properties

POPs are hydrophobic and lipophilic, that is, they are fat-soluble while resisting breakdown in water. Their lipophilic tendency enables them to concentrate in fatty tissues of organisms and bioaccumulate in the food chain. As noted by Eckley (2001: 28), the levels of POPs detected in organisms that are high on the trophic levels, such as seals, polar bears, predatory birds, mammals, and humans, are sometimes thousands of times higher than the levels found in those in the immediate surroundings (also see Wania and Mackay, 1999). Biomagnification is an increase in the concentration of POPs and other organochlorinated chemicals in organisms as they pass through the food chain.

The Grasshopper Syndrome

POPs are known to be highly mobile, traveling long distances even to the remote corner of the Earth (see Koziol and Pudykiewicz, 2001; Eckley, 2001; Wania and Mackay, 1996). They exhibit a process known as the "grasshopper effect," in which these chemicals go

Table 5.1 Top-Priority POPs, Uses, and Their Adverse Health Effects

Class	Chemical	Uses	Half-Life in Soil (years)	Adverse Health Effects
A. Agricultural and Landscape Chemicals: Pesticides	1. Aldrin	Insecticide	N/A	Carcinogenic, malaise, dizziness, and nausea
	2. Chlordane	Insect and termite control	1	Carcinogenic
	3. DDT	Insecticide	10–15	Cancer of liver, immune system suppression
	4. Dieldrin	Insecticide	5	Liver and biliary cancer
	5. Endrin	Insecticide	up to 12	Cancers
	6. Heptachlor	Insect and termite control	up to 2	Cancers, mutations, stillbirths, birth defects, liver disease
	7. Hexachloro-benzene (HCB)	Fungicide	2.7–22.9	Cancers, mutations, birth defects, fetal and embryo toxicity, nervous disorder, liver disease
	8. Mirex	Insecticide, termicide	up to 10	Acute toxicity, possible cancers
	9. Toxaphene	Insecticide	3 months	Carcinogenic, chromosome aberrations, liver and kidney problem
B. Industrial Chemicals	10. Poly-chorinated biphenyl (PCB)	Industry manufacture, co-planar	10 days to 1.5 years	Cancers, mutations, birth defects, fetal and embryo toxicity, neurological disorder, and liver damage
	11. Dioxins	By-product	10–12	Peripheral neuropathies, fatigue, depression, liver disease, embryo toxicity
	12. Furans	By-product	10–12	Peripheral neuropathies, embryo toxicity, liver problems

Source: Adapted from Epstein, S. S., L. O. Brown, and C. Pope. *Hazardous Waste in America*. San Francisco, CA: Sierra Club Books, 415–427; UNEP. Report of the Intergovernmental Negotiating Committee for an International Legally Binding Instrument for Implementing International Action on Certain Persistent Organic Pollutants on the Work of Its Fifth Session, Geneva: UNEP/POPS/INC. 5–7, (December 26, 2000); The World Bank and CIDA. *Persistent Organic Pollutants and the Stockholm Convention: A Resource Guide*. Washington, DC: CIDA, 2001.

through cycles of volatilizations and condensations, that is, evaporation and atmospheric cycling in warmer climates and condensation and deposition in colder climates, thus moving these chemicals to remote regions where they have never been produced or used. Colborn et al. (1996: 106) state:

> These synthetic chemicals move everywhere, even through the placental barrier and into the womb, exposing the unborn during the most vulnerable stages of development.. . . When a new mother breastfeeds her baby, she is giving him/her more than love and nourishment—she is passing on high doses of persistent chemicals as well.

POPs tend to reach their highest level of concentrations in the cooler regions of the globe. The indigenous people of the Arctic who depend on a traditional diet composed of foods with high fat content are especially at risk of POP contamination and the resultant adverse health effects. Virtually all living organisms in any part of the globe now carry detectable levels of POPs in their tissues. As stated earlier, there is a growing evidence of POP contamination in our foods, in human blood, and in breast milk (Colborn et al., 1996; Thornton, 2000; Schafer et al., 2001). Even though establishing a direct one-on-one cause and effect of a specific xenobiotic and the resultant adverse health conditions is contentious among experts, scientists, and lay people, some major health problems associated with POPs and other related toxic chemicals have been documented in the literature (see Colborn et al., 1996; Thornton, 2000; Solomon and Schettler, 1999; WWF, 1999; Eckley, 2001; Smith, 2001; Myers, 2002). The major health impacts of POPs are discussed in the following section.

Exposure Pathways and Health Issues Linked with POPs

Given their ubiquity and persistence in the environment, there is no safe place for escaping POP contamination. Typical routes of exposure include workplace (in agriculture and industries), dietary exposure, and direct contact with contaminants in the air, buildings, water, lawns, parks, and soil, including, but not limited to, accidental releases (Lallas, 2000/2001, 2002). Pervasive harm to both wildlife and humans by POPs have been documented extensively in the literature. In the case of the former, adverse effects of POPs and related toxic chemicals range from egg-shell aberration in birds to extinction of certain bird species (Carson, 1962; Colborn et al., 1996;

WWF, 1999); other serious effects include cancers, twisted spines and skeletal deformations, and death of beluga whales. In Florida's Lake Apoka, stunted penis, hormone disruption, and reproductive failure have been found among alligators, and disrupted reproductive development, deformity, immunotoxicity, hormonal deficiencies, and overall population decimation have also been reported (Abelsohn et al., 2002; Jones and de Voogt, 1999; WWF, 1999; Swan et al., 1997).

In the case of humans, the litany of health problems related to POP contamination is quite extensive including allergies, birth defects, cancers, embryo toxicity, endocrine (hormone) disruptions, decreased sperm count, diabetes, hypersensitivity, hypospadias (an arrested development of the urethra, foreskin, and ventral aspect of the penis), kidney and liver dysfunctions, learning and behavioral problems especially among children, mutations, nervous disorders, premature births, and stillbirths (Hauser et al., 2002; Swan et al., 2000; Guillette et al., 1998; Baskin et al., 2001; Thornton, 2000). Children and infants are particularly more vulnerable than adults to the adverse effects of POPs. Wargo (1996: 11) notes that children and infants seem to be especially vulnerable to carcinogens during periods when their cells are normally developing most rapidly, generally between conception and age five.

The last column of table 5.1 shows a partial list of the specific adverse health effects of selected POPs.[3] In a review of 101 studies published between 1934 and 1996, Swan et al. (2000: 964) note the basic finding of a 50 percent decline in sperm count among US/Canadian, European/Australian men, but not among non-Western men. Similar findings of an association between certain POPs, such as PCBs and p,p'-DDE, and abnormal sperm count, motility, and morphology, among 29 subjects recruited from the Massachusetts General Hospital Andrology Laboratory, was recently reported in the journal *Environmental Health Perspectives*, by Hauser et al. (2002). Thus, empirical evidence links specific POPs to male reproductive dysfunctions.

An evaluation of POP residue data from several reputable sources produced some startling findings. In the US Food and Drug Administration (FDA) residue monitoring in 1999, POP residues were detected in several food items grown and consumed locally as well those imported from abroad to the United States. Table 5.2 shows the top 10 foods mostly contaminated by specific POPs.[4] Residues were found in virtually all food categories including baked goods, fruits, vegetables, meat, poultry, and dairy products. The exposure per day was highest in the

Table 5.2 Top 10 Foods Most Contaminated with Specific POPs(*)

Items	Selected Food				POP Chemicals				
	Chlordane	DDE	DDT	Dieldrin	Dioxin	Endrin	Heptachloro	HCB	Toxaphene
Butter	(*)	(*)	(*)	(*)	(*)		(*)	(*)	
Cantaloupe		(*)		(*)	(*)		(*)		(*)
Cucumbers/Pickles	(*)	(*)		(*)		(*)	(*)		(*)
Meatloaf		(*)		(*)	(*)		(*)	(*)	(*)
Peanuts		(*)		(*)				(*)	(*)
Popcorn(*)				(*)					(*)
Radishes(*)	(*)	(*)	(*)	(*)		(*)	(*)		(*)
Spinach		(*)	(*)	(*)					(*)
Summer Squash	(*)	(*)		(*)		(*)		(*)	(*)
Winter Squash	(*)	(*)		(*)		(*)		(*)	(*)

Sources: US Food and Drug Administration. Total Diet Study, September 2000, online at: http://vm.cfsan.fda.gov/~dms/pes99rep.html; Schecter, A., J. Startin, C. Wright, M. Kelly, O. Papke, A. Lis, M. Ball, and J. R. Olsen. "Congener-Specific Levels of Dioxins and Dibenzofurans in US Food and Estimated Daily Dioxin Toxic Equivalent Intake." *Environmental Health Perspectives* 102 (1994): 962–966; US EPA, "Estimating Exposure to Dioxin-Like Compounds." *Volume II: Properties, Sources, Occurrence and Background Exposures.* Washington, DC: U.S. EPA Office of Research and Development, 1994.

Table 5.3   POPS Pesticides Found in Total Diet Study in 1999

Frequency of Occurrence of Pesticide Residues

Pesticide	Total No. of Findings	% Occurrence
DDT*	225	22
Chlorpyrifos-methyl	188	18
Malathion	175	17
Endosulfan	151	15
Dieldrin*	145	14
Chlorpyrifos	93	9
Chlorpropham	70	7
Permethrin	54	5
Iprodione	48	5
Chlordane*	36	3
Heptachlor*	36	3
Lindane	33	3
Thiabendazole	33	3
BHC, alpha+beta+delta	32	3
Hexachlorobenzene (HCB)*	32	3

Note: Items asterisked are POPs recently banned by the Stockholm International treaty.

Source: US Food and Drug Administration. Pesticide Program, Residue Monitoring 1999. Washington, DC: FDA.

southeast of the United States (70) and lowest in the Midwest (63). Table 5.3 presents the number of food items analyzed in the Total Diet Study conducted by the FDA and the findings of POPs and pesticide residues in terms of number and percentage of occurrence.[5] These findings seem instrumental to the Bush administration support of the Stockholm Convention on Persistent Organic Pollutants.

Beginning in 2000, the EPA has established more stringent Toxics Release Inventory (TRI) reporting benchmarks for persistent bioaccumulative toxic chemicals (PBTCs) originally on, or recently added to, the TRI chemical list. The TRI PBTCs of concern that exhibit the properties described above include dioxin and dioxin-like compounds, polycyclic aromatic compounds (PACs), PCBs, certain pesticides (all members of the POP family), and some inorganic elements including lead and lead compounds, and mercury and mercury compounds among others. In the United States, PBTCs accounted for 12 percent of reported on- and off-site toxic releases in 2007 (EPA, 2009). As mentioned earlier, POPs and PBTCs are of particular concern for environmental health because in addition to being very toxic, they persist in the environment for long periods of time, they are not readily biodegradable, and they bioaccumulate in body tissue.

THE PRECAUTIONARY PRINCIPLE AND
INTERNATIONAL AGREEMENT ON POPs

In the United States and other industrialized nations, toxic chemicals are being produced and released at a pace faster than the enactment of laws that are supposed to regulate them (Adeola, 2002). Many of the new chemicals are persistent, deadly, and transboundary in nature, requiring international laws and cooperation to regulate them. The precautionary principle is now being advocated in the European Union (EU), the United States, and across the globe. This principle suggests that whenever there is scientific uncertainty about the safety or potentially serious harm from chemicals or technologies, manufacturers or decision makers shall do everything possible to prevent harm to humans and the environment. It stipulates that when an activity raises threats of harm to human health or the environment, precautionary measures should be implemented even if some "cause-and-effect" relationships have not been fully established (Raffensperger, 2003: 4). In other words, "it is better to be safe than to be sorry," and manufacturers of toxic chemicals should be held accountable for any serious adverse health effects of these chemicals to humans and the environment. Environmentalists and other scholars interpret this principle to suggest that if any uncertainty exists about the safety of a technology, then it ought to be strictly restricted or banned. Whenever technologies have the potential to cause serious, large-scale, and permanent damage to health and the environment, we should look to the future and preempt possible injury, not only the damage that has already taken place (Thornton, 2000). Absolute safety to humans and the environment is required of any technology under the precautionary principle (see Goklany, 2001).

The aspiration of many advocates of the precautionary principle (a restatement of the popular rendition of the Hippocratic Oath, "First do no harm") is to have it become a cornerstone for constructing public health and environmental policies. From a critical perspective, Goklany (2001) argues that taking an absolutist stance in the presence of uncertainty as required by the precautionary principle may also produce unintended consequences. Attempting to solve one problem through a restriction or a complete ban of a given technology in the absence of a substitute may create more serious problems as is the case with DDT in developing countries. It is especially important to make sure that policies derived from the precautionary principle are not counterproductive for public health and the environment. Thus, universal standardized policies and regulations of certain POPs may be counterintuitive.

To make sure that good intentions do not yield unintended negative outcomes, Goklany (2001: 8–10) offers a framework that would allow the precautionary principle to be used in situations where the overall result might be ambiguous in terms of positive and negative impacts of a policy choice. The framework consists of a set of hierarchical criteria that can be employed to rank the various threats raised or diminished by a given policy based on the nature, magnitude, immediacy, uncertainty, and persistence of each threat, and the extent to which it can be minimized or eliminated. This framework is anthropocentric in the sense that threats to human health and well-being take precedence over threats to the environment and nonhuman species. More specifically, the following criteria are suggested within the framework:

1. The Public Health Criterion. This criterion suggests that morbidity and mortality threats to human beings should outweigh similar threats to members of other species, and other nonmortal threats to human health should be given priority over threats to the environment (with exceptions based on the nature, severity, and magnitude of the threat).
2. The Acute vs. Chronic Threat Criterion. Immediate attention should be given to acute threats over threats that could occur in the distant future.
3. The Uncertainty Criterion. This implies that threats of harm with higher probabilities of occurrence should take precedence over those with lower probabilities.
4. The Expectation-Value Criterion. When confronted with threats that are equally certain, precedence should be accorded those that have a higher expectation value. For instance, an action yielding fewer expected casualties should be preferred over the one producing mass casualties.
5. The Adaptation Criterion. If technologies are available to address or manage the adverse outcomes of an impact, then the impact can be discounted inasmuch as people are able to cope and adapt to the situation.
6. The Irreversibility Criterion. Greater priority should be given to outcomes that are persistent and irreversible.

Other scholars have suggested that implementing precautionary measures that impose more stringent requirements on old and new chemical products is absolutely imperative. The following criteria recognized by the Swedish Chemicals Policy Committee have been suggested in the literature (Myers, 2002: 4; Fredholm, 2000):

1. Complete elimination of persistent bioaccumulative compounds even without demonstrating their toxicological risks.
2. Removal of endocrine-disrupting compounds from consumer products and phasing out their environmental release.
3. Introduction of a new approach in which safety is assured beyond any reasonable doubt prior to the introduction, distribution, and mass use of toxic chemical products.

Other Principles for Reducing Toxics

In addition to the precautionary principle, at least two other principles for reducing global toxic load in the environment have been proposed in the literature, including the principles of *reverse onus* and *the least toxic alternative*. As explained above, the idea that both private and public interests should act to prevent harm before it occurs is the essence of *the precautionary principle*. For *the principle of reverse onus*, it is safety, rather than harm, that should warrant demonstration; that is, the burden of proof is shifted from the public to those who produce or manufacture, sell, import, or use the toxic substance in question. Thus, this principle requires those who seek to introduce new chemicals into the environment to first demonstrate that what they are proposing to do will cause no harm to anyone in the environment. The least toxic alternative principle stipulates that with all activities having potential public health consequences, toxic substances will not be used when nontoxic alternatives are available to accomplish the same task. This implies using the least harmful method of solving problems, whether it is pest control, weed control, or water treatment. According to Steingraber (2000: 33), the principle of least toxic alternative would move us away from protracted, unwinnable debates over how to quantify the cancer risks from each carcinogen released into the environment, and where to set the legal limits for their presence in air, water, home, workplace, and in consumer products. These principles serve as important policy instruments for public administrators interested in protecting both the environment and public health while promoting sustainable society.

The Stockholm Convention on POPs

At a conference in Stockholm, Sweden, in May 2001, the international community adopted the new Stockholm Convention on POPs. At that time, more than 90 countries signed the convention and Canada was the first to ratify it. This convention has been signed by more than 122 countries as of March 18, 2002, and it is based upon the precautionary principle, which reflects some of the criteria mentioned earlier. Specifically, the convention is designed to protect human health and the environment from POPs, which have become a prominent global problem. The convention asks its member countries to ban the 12 POPs mentioned earlier and eliminate or restrict the use of other toxic chemicals that are proven to have the properties of POPs. In addition, members are required to take strong measures to prevent or control the release of certain POPs formed as by-products of various industrial combustion activities, and to ensure the safe and proper disposal of such substances when they become waste. Furthermore, provisions were made to add new chemicals to the list of banned POPs. This convention was structured to address POPs that are intentionally manufactured, such as pesticides, insecticides, rodenticides, and fungicides, whose use is restricted to disease vector control (e.g., DDT for controlling mosquitoes causing malaria), and those, such as dioxins, furans, PCBs, and HCB, that are unintentionally produced and released by accident due to human activities (The World Bank and CIDA, 2001). Countries are required to make determined efforts to identify, label, and remove PCB-containing equipment from use by 2025, and manage those wastes in an environmentally sound manner no later than 2028. Stockpiles and wastes containing POPs must be managed and disposed of in a safe, efficient, and environmentally sound manner, taking into account international norms, standards, and guidelines. Also, to ensure implementation of various articles of the convention, provisions were made for information exchange and increasing public awareness and education about the adverse health effects of POPs.

Chapter Summary

This chapter covers the relationship between environmental pollution and health problems affecting human and nonhuman species. Specifically, it reviewed existing evidence on POPs—a class of synthetic organochlorine chemicals and products introduced after World War II—and their adverse health effects on society and wildlife. Their fundamental characteristics including toxicity, persistence, ability to migrate long distances, and bioaccumulation within the food chain are discussed. The reactions of the international community, especially the United Nations (UN) Stockholm Convention on POPs signed by 122 countries are also discussed. The future needs of substitutes to harmful chemicals and mitigation of health problems already caused are addressed as well. The use of the precautionary principle, the principles of reverse onus, and the least toxic alternative as a guide to public health and environmental policies is emphasized.

As a result of accelerated production, use, and release of heterogeneous synthetic organochlorine chemicals over the last 60 years, humans and wildlife are now struggling with multiple environmental contaminants. POPs represent the major culprit for most pollutants and the health problems reviewed in this chapter. But our knowledge is still limited about other chemicals, and the synergistic effects of POPs are still under scientific investigation. Even though many synthetic pesticides and organochlorine products have brought better things to life in terms of increased food production and availability, disease control, aesthetically appealing lawns and gardens, and other conveniences in everyday lives, paradoxically, these toxic chemicals are now frightening as they represent the very essence of dread in terms of the havoc they wreak on the environment, the wildlife, and human health across the globe. Their ability to travel long distances to places where they have never been produced or used begs international cooperation efforts to address these transboundary challenges.

Precautionary measures are required at every level from the individual to the global to eliminate these elixirs of death and find environmentally friendly substitutes. The Stockholm Convention on POPs may be considered the most significant international initiative designed to mitigate harm to human health and the environment from POPs and related chemicals. Thus, it represents a step in the right direction. Among the 12 POPs, DDT is the only one not completely banned, especially in the Third World, where its application to control malaria is paramount to any adverse health considerations. Despite remarkable achievements in reducing the production, distribution, and application of POPs, these toxic pollutants are now ubiquitous in the environment as evident by their presence in foods from around the world and in breast milk of nursing mothers (Schafer et al., 2001; Abelsohn et al., 2002). Even though all the 12 POPs have been banned or phased out in advanced industrial nations including the United States, POP residues in agricultural products from underdeveloped regions continue to represent a major pathway for human exposure.

Consistent with the cultural lag theory, thousands of toxic and hazardous chemical products are developed and released to the environment on a regular basis; unfortunately the laws designed to regulate these products are decades behind (Adeola, 2002). Furthermore, funding for the development of innovative technologies that carry significant risk to health and the environment generally overwhelmed that for the research to understand the magnitude, distribution, toxicity, and adverse impacts of chemical contamination (see Wargo, 1996). Thus, reversing these previous trends is an important aspect of arresting the risks posed by POPs and allied hazardous products. Rigorous scientific endeavors to develop safer, effective, and nonpersistent substitutes and mitigation of adverse health effects of all POPs and chemicals with similar properties are strongly encouraged.

PART III

CONTAMINATED COMMUNITIES AND REGULATORY RESPONSES

Intentional or unintentional releases of hazardous materials to the environment often result in the exposure of innocent populations to health risks; and in some cases, contamination of an entire community is the outcome. Similarly, industrial disasters have a tendency to kill, maim, or inflict chronic physical, emotional, and psychosocial injuries on the victims. Part III of this book is devoted on the one hand to the problems of hazardous waste contamination of communities and human bodies, and on the other hand to the technological industrial disasters also wrecking havoc on the surrounding vulnerable populations and communities along disaster-prone locations. In most cases, regulations are introduced ex post facto. Two chapters in this part explore the case studies of several communities both within the United States and cross-nationally that have experienced either toxic waste contamination or contamination brought about by industrial accidents—and the subsequent legislative responses.

Chapter 6 covers selected cases of contaminated communities—including Love Canal, New York; Woburn, Massachusetts; Agricultural Street Landfill, New Orleans; Seveso, Italy, dioxin contamination; Koko, Nigeria; and Bhopal, India. Community disruption, conflict, and prolonged struggle are typical of these communities. Chapter 7 presents the regulatory frameworks and specific statutes, norms, or international agreements aimed at curbing or controlling toxic waste flow and industrial accidents. Even though disasters are ubiquitous and inevitable in complex advanced industrialized society, precautionary measures are absolutely imperative for preventing such events from manifestation. Key legislations within the United States, the European Union (EU), and international accords are discussed.

6

COMMUNITIES CONTAMINATED BY TOXIC WASTES AND INDUSTRIAL DISASTERS: SELECTED CASES

INTRODUCTION

Industrial disasters, reckless disposal of toxic wastes by firms, and misguided housing developments in risky landscapes are inextricably associated with increased incidence of community contamination by toxic or hazardous substances in recent decades.[1] According to the 2007 World Disaster Report, the total number of both technological and natural disasters has increased markedly from the 1990s to 2006 (International Federation of Red Cross and Red Crescent Societies, 2007). Edelstein (1988) has described "toxic exposure" of communities as a modern plague with enduring consequences. He coined the term "contaminated community," defined as any residential area or human settlement located within *or in proximity* to the designated boundaries of a known environmental hazard exposure.

Accidental releases of toxic chemical compounds and hazardous waste dump sites are among the major sources of community contamination. According to the Agency for Toxic Substances and Disease Registry (ATSDR), the annual cost of just four childhood health problems—asthma, cancer, lead toxicity, and developmental disabilities—associated with toxic chemical exposure in the community, exceeds 54 billion dollars. Most toxic disasters are anthropogenic or technological in origin. Both the social relationship and the ecological balance are disrupted by a toxic contamination event. A corrosive community typically evolved in the course of intensification of conflict between the victims and the party responsible for the disaster (Erickson, 1976, 1994). The metaphor of toxic waste as the plague of the twentieth and the twenty-first centuries is quite appropriate for the growing number of contaminated communities within the United States and around the world since the end of World War II.

Mitchell (1996) makes a distinction between two major types of technological/anthropogenic disasters as "routine" disasters and "surprises." The former are clearly understood by experts and are manageable by relying on long-established principles and procedures, while the latter tend to confound both expert and lay expectations as they are very different and less well understood. Examples of surprises are the Union Carbide disaster in Bhopal, India, the Chernobyl nuclear meltdown in the former Soviet Union (now Ukraine), the Katrina natural-technological (nat-tech) flood in New Orleans, and the Deep Water Horizon oil rig disaster spewing uncontrollable amounts of crude oil into the Gulf of Mexico along the Louisiana Gulf Coast communities in 2010. Some of these surprises are discussed in this chapter. The cases of Love Canal, New York; Woburn, Massachusetts; Agriculture Street in the 9th Ward of New Orleans; Union Carbide disaster in Bhopal, India; Seveso dioxin disaster; and toxic waste dumping and contamination of Koko village in Nigeria are particularly instructive relative to hundreds of other community contamination cases. Each of the cases is briefly discussed in the following sections.

THE LOVE CANAL

Levine (1982) provides a detailed account of the contamination episode at Love Canal, New York.[2] The history of Love Canal began in 1892, when William T. Love proposed to build an eight-mile power-generating canal connecting the Niagara River to Lake Ontario and Lake Erie. Among Love's objectives were to develop a navigable channel along the Niagara and to provide cheap electricity based on direct current (DC) to the area. Most importantly, his plan was to develop a "model city" of about 600,000 people. As noted by Epstein et al. (1982: 91), in Love's envisioned model city, industries would be attracted to the shores of the canal to capitalize on cheap and abundant hydroelectric energy. The canal was dug halfway (approximately 1 mile or 1.6 km) when the project was abandoned due to a number of factors including the introduction of alternating current (AC) by Nikola Tesla, which could transmit electricity over a longer distance at a lower cost than the DC at the core of Love's plan. Also, the economic depression of the 1890s led to substantial erosion of financial support for the project, and bankruptcy was inevitable.

Love abandoned the project and left a partially excavated canal 100 feet wide, 10 feet deep, and 3,000 feet long. By the 1920s and 1930s, the canal was filled with water and used as a swimming hole by local residents. In 1942, the Hooker Electro-Chemical Corporation (later known as Hooker Chemical and Plastics Corporation and a subsidiary of Occidental Petroleum) secured permission from the Niagara Power and Development Corporation to dump chemical wastes in the canal. The city of Niagara Falls and the US Army also used the site for dumping solid wastes and chemical waste and parts of the Manhattan project. By 1946, however, Hooker acquired the entire Love Canal from the Niagara Power and Development Corporation and used it exclusively for dumping toxic chemical wastes. The canal, embedded in an area of orchards and farms, was considered an excellent waste disposal site. As was common during the period, the company did not install any liner inside the canal to prevent leaching. About 21,000 metric tons (or 43 million pounds) of heterogeneous toxic chemical wastes—including alkalis, caustic soda, dioxin, fly ash, mercaptans, sulfides, polychlorinated hydrocarbons, fatty acids, perfumes, trichlorophenols (TCPs), solvents for rubber, synthetic resins, and other by-products of dye—were deposited as liquids and solids in metal drums and other types of containers (Levin, 1982: 9, Blum, 2008; Phillips et al., 2007). Over 13 million pounds of benzene hexachloride, more than 4 million pounds of chlorobenzene, and about half a million pounds of TCP were deposited in the canal. More than 200 different varieties of chemicals and chemical compounds were identified to be present at the site (Phillips et al., 2007). By 1953, the canal was full to capacity and was covered up with clay soil and dirt. As mentioned by Levin (1982: 7), what took place next sets in motion events that would affect thousands of families for several years, and perhaps for the rest of their existence.

In 1953, the Hooker Chemical Company (later acquired by OxyChem in 1968) sold the 16-acre land including the land-filled canal to the City of Niagara Falls Board of Education for a token or symbolic fee of one dollar ($1) with a declaration that the canal was filled with chemical waste products and a disclaimer of any future liabilities. However, there was no public health warning about the potential adverse health effects of the wastes. Shortly after the acquisition of the land, a school was built and the rest of the land was sold for development involving the construction of single-family homes. Home and street construction directly adjacent to the canal was accelerated by the mid-1950s.

The first sign of health problems occurred in 1958, when school children received serious chemical burns from leaking drums while playing around the road construction sites. The leaking drums were covered up with dirt as a quick fix; land subsidence in the school playground occurred periodically and the holes were filled with soil. In 1976, residents of single-family dwellings adjacent to the Canal started to complain about chemical odors and seepage of wastes from the Love Canal. In the spring of that year, a record-breaking snow-melt occurred and forced toxic chemicals from below the earth to the surface of the Love Canal neighborhood and into the residential structures and yards of families living near the elementary school. Among several signs of trouble were malodorous fumes in base-ments, minor explosions, cracked foundations of fairly new homes, chemical residues showing up inside basements and pools, and water pumps that were clogged and corroded from the hard chemical sub-stances, all of which became quite upsetting to the people.[3] Eckard C. Beck, EPA regional administrator, stated in the January issue of *EPA Journal* in 1979:

> Love Canal is one of the most appalling environmental tragedies in American history. . . ..I visited the canal area at that time. Corroding waste-disposal drums could be seen breaking up through the grounds of backyards. Trees and gardens were turning black and dying. One entire swimming pool had been popped up from its foundation, afloat now on a small sea of chemicals. Everywhere the air had a faint, chok-ing smell. Children returned from play with burns on their hands and faces. And then there were the birth defects.[4]

Among the potentially exposed population were several hundred residents living within one block of the toxic waste dump and another 3,000 residents living within four blocks of the canal. By 1978, Love Canal had gained national (and perhaps international) spotlight as a case of a contaminated community. At this time, over 400 toxic chemicals most of which are carcinogenic, teratogenic, and mutagenic were identified at the site. Residents began to understand the roots of their ill-health or the somatic conditions ranging from untimely deaths of relatives, cancers, miscarriages, birth defects, skin rashes and irritations, to major organ dysfunctions (see Levin, 1982; Miller and Miller, 1991). Lois Gibbs was prominent among the residents of the community who organized grassroots efforts and conducted popular epidemiology with findings confirming toxic contamination and human health problems, including that of her own son who suf-fered from epilepsy and several other adverse health conditions linked

20 Years ago Love
Canal families
thought they were
safe after
evacuating their
contaminated
neighborhoods...

...only to discover they
couldn't protect their
families from chemical
exposures anywhere.
Love Canal could be
evacuated but...

We can't Evacuate

Figure 6.1   Poster of Lois and Missy Gibbs Promoting Environmental Justice.
Source: Ecumenical Task Force, Courtesy of State University of New York at Buffalo.

to exposure to toxic chemicals in the vicinity (see figure 6.1). She later
discovered that other families with children in her neighborhood have
similar experiences as numerous children suffered from unexplained
illnesses ranging from skin disorder to retardation.

   After carefully reviewing the initial scientific studies of toxicity
of Love Canal and in response to the public outcry of the residents,
Robert Whalen, the health commissioner for the state of New York,
issued an official announcement on August 2, 1978, that the residents in proximity to the canal presented evidence of higher-than-
normal levels of spontaneous abortions, birth defects, and that at least

one human carcinogen had been found within the vicinity. As a declaration of "the state of emergency" at Love Canal, Whalen further stated:

> The Love Canal Chemical Waste Landfill constitutes a public nuisance and extremely serious threat and danger to the health, safety, and welfare of those using it, living near it, or exposed to the conditions emanating from it, consisting, among other things, of chemical wastes lying exposed on the surface in numerous places and pervasive, pernicious, and obnoxious chemical vapors and fumes affecting both the ambient air and the homes of certain residents living near such sites. (Levin, 1982: 28; Blum, 2008: 27)

In addition to physical health problems among the residents, there were social, economic, and psychological dimensions to the contamination episode at Love Canal, as homes became worthless, and a major source of dread and anxiety about the future (see figure 6.2) prevailed.

Love Canal was declared a Federal Disaster Area in 1978. In August of 1978, Governor Hugh Carey ordered that 236 families be

Figure 6.2 Aerial Photograph of Abandoned Contaminated Residential Structures Adjacent to Love Canal Site Displaying Toxic Waste Rising to Ground Surface.
Source: Courtesy of State University of New York at Buffalo Archives.

relocated out of Love Canal. Subsequently, President Jimmy Carter gave an order for the relocation of about 800 homeowners at a cost of more than $30 million; thus, the federal government assumed the responsibility of buying out about 800 homes in addition to cleaning up the toxic wastes at Love Canal. Clearly, there was a major community destruction, social disruption, and loss of social capital in addition to different levels of conflict caused by the Love Canal toxic contamination. By 1990, the EPA has issued its Record of Decision (ROD) indicating that the danger of toxic chemicals was no longer a threat in most of the Love Canal area. As noted by Phillips et al. (2007: 318), numerous cleanup activities—encompassing landfill containment, leachate collection and treatment, and the removal and disposition of the containment sewer and creek sediments and other wastes—were carried out at the site. These activities supposedly eliminated the major contamination exposure pathways, making the site safe for the nearby residents and the environment; and by early 1994, it was announced that the cleanup of the condemned homes in the Love Canal was over, and it was safe to move back to the area. The real estate company offering the decontaminated and refurbished homes in Love Canal for sale changed the name of the area to "Sunrise City."

Nevertheless, many environmental and community activists remained skeptical about the EPA's decision and generally unsatisfied with the agency's ROD. Future morbidity and premature mortality among the exposed residents of the former Love Canal were of major concern among environmental health researchers, toxicologists, and epidemiologists (see Gensburg et al., 2009). As indicated by Clapp (2009), while the Love Canal saga served as a "warning signal" for several communities hosting such locally undesirable land uses (LULUs) across the country, this story remains open-ended or unfinished.

As mentioned before, conflict is endemic in cases of toxic contamination of communities. The litigation against Hooker Chemical Company took several years to be resolved. After a protracted court battle, in 1988 (10 years after the event), Judge John Curtin found Occidental Petroleum Corporation (successor of Hooker) responsible for the environmental contamination and ordered the company to pay an amount of $98 million as restitution to the state of New York. The US federal government later settled with the company for $129 million in the same year. The lawsuits filed by individual residents were eventually resolved as well by 1997 for over $20 million (see Blum, 2008; Phillips et al., 2007). In 1999, OxyChem also agreed to repay the federal government and the state of New York the sum of $7.1 million for the cleanup of the Love Canal. Furthermore, the company assumed the responsibility for future treatment of wastes at the site. One key development associated with the Love Canal contamination episode was the US Congress enactment of the Comprehensive Environmental Response, Compensation and Liability Act (CERCLA) of 1980, otherwise known as the "Superfund" law. This legislation deals with environmental response and provides the means for dealing with emergency situations and chronic toxic material releases. While CERCLA establishes procedures to prevent and mitigate environmental problems, it also sets up the system for compensating the victims of toxic contamination and appropriate liability for the responsible party. Furthermore, the trust fund established by this legislation was designed to pay for cleanup and revitalization of the most contaminated sites across the United States, and its amendments and authorization in 1986 established the ATSDR.

WOBURN, MASSACHUSETTS

In Woburn, Massachusetts, chemical companies, tanneries, and glue-manufacturing plants dumped toxic wastes into empty lots for over a century, creating approximately over 60 acres of a toxic brew of arsenic, lead, chromium, and synthetic organic compounds. Through "mid-night dumping" operations, 184 drums (55 gallons each) of toxic chemicals were deposited in a vacant lot near the east bank of the Aberjona River. Residents obtain their water supply from several wells in the area. Although the state environmental agency found that the drums had not leached toxic chemicals into the wells, tests revealed that wells G and H from which most residents obtained their water, were highly contaminated with chlorinated organic compounds including trichloroethylene (TCE) and tetrachloroethylene (also known as perchloroethylene (PCE)). Soils at the site also contained volatile organic compounds (VOCs), PAHs, PCBs, pesticides, and high concentrations of heavy metals including arsenic, chromium, lead, mercury, and zinc (US EPA, 1993; Bair and Metheny, 2002). The unpleasant results were acute and chronic toxicity upon exposure leading to death, leukemia, and various other debilitating health problems, especially among the children (see Brown and Mikkelson, 1990; Freudenberg, 1984: 31). Other health problems associated with the contamination of wells G and H included cardiac arrhythmias and liver, nervous system, and immune system disorders (Kennedy, 1997).[5]

This contamination episode led to the formation of a grassroots group and a subsequent long legal battle against W. R. Grace and Company, Beatrice Foods Company, and UniFirst Corporation alleged to be the potentially responsible parties (PRPs) for the toxic contamination in Woburn (see Brown and Mikkelsen, 1990; Kennedy, 1999). UniFirst settled for $1.05 million without admitting fault; Grace and Company settled out of court for an estimated sum of $8 million; and the case against Beatrice Foods, Inc. was dismissed. Recently, in 2009, the EPA has indicated its intention to clean the groundwater in East Woburn and restore it to drinking-water quality (U.S. EPA, 2009). The cleanup cost is estimated at about $70 million (to be paid by the PRP), and it would take between 30 and 50 years to be completed.

THE AGRICULTURE STREET LANDFILL (ASL) COMMUNITY IN NEW ORLEANS

Agriculture Street is a 95-acre Superfund site in New Orleans East, housing a community of 900 people who are predominantly African Americans with an average income of $25,000 per year. The history of Agriculture Street Landfill (ASL) dates back to 1909 when the area was first designated as the city dump site (Finch, 1994). It operated as a municipal landfill from 1910 through the late 1950s, when it was closed. In 1965, ASL was reopened for about one year as an open burning and disposal depot for all debris and wastes from Hurricane Betsy. A wide variety of wastes were dumped at the site without specific standards, regulations, or guidelines about the type of wastes to be accepted. In addition to municipal solid wastes, pesticides, petrochemical wastes, tires, old lead-based paints, ashes, and solvents were deposited and buried at the site. Similar to Love Canal, redevelopment of the site was undertaken in 1969 with the construction of low-to-middle income homes (Press Park and Gordon Plaza), condominiums, apartment complexes, and an elementary school directly on top of the old ASL. Between 1986 and 1993, there were conflicting claims about the extent of contamination and associated health problems in the community. About 48 acres of the landfill remain undeveloped.

Scientific tests conducted revealed the presence of more than 100 toxic chemicals in the soil of the community, some of which are known carcinogens, including PCBs, polynuclear aromatic hydrocarbons (PAHs), lead, arsenic, chlordane, and DDT (see figure 6.3). In 1994, Agriculture Street was declared a National Priority List (NPL)

Figure 6.3 Photograph of a Resident of Agriculture Street NPL Site Displaying a Sign Showing the Extent of Contamination and a Demand for Justice.

site by the EPA. Unlike the Love Canal, however, despite media coverage, grassroots mobilization efforts, and residents' pleas, relocation of residents and buyout of contaminated houses were not favored by the federal government. The preferred remedial action by the EPA for the entire community was the removal and replacement of the top soil. Most residents feel strongly that they are victims of environmental injustice and environmental racism (Adeola, 2000a). Nevertheless, the EPA issued its ROD not to implement any remedial measure for the elementary school and groundwater beneath the site. It was acknowledged that the "no action" remedies will result in hazardous substances remaining within the site; however, a review will be conducted every five years consistent with the CERCLA provisions. Meanwhile, the soil under building structures, homes, streets, and sidewalks remained severely contaminated. The EPA gave homeowners certificates indicating that their property had been partially remediated. Also, a list of restrictions on the use of their property was provided.

Many Agriculture Street homeowners, community activists, and environmental justice advocates remained skeptical about the efficacy of the remedial measure undertaken by the EPA. As noted in environmental justice literature, predominantly African American communities with NPL sites are less likely to be given a comprehensive, total remedial, or relocation option by the EPA relative to a predominantly white community. Numerous adverse health conditions, economic, social, and psychological problems have been reported by Agriculture Street residents (Adeola, 2000a).

Similar to Love Canal, the Concerned Citizens of Agriculture Street Landfill (CCASL) was formed as a grassroots organization involved in the struggles over ill-health, injustice, and demands for environmental justice through adequate remedial measures. Dissatisfied with the EPA's ROD, the residents filed a class action lawsuit against the city of New Orleans and the Housing Authority of New Orleans (HANO) for damages and relocation costs. The class action lawsuit was originally filed in 1993 by the community attorney. Bullard (2008) has indicated that the residents of Agriculture Street filed the lawsuit in order to force their relocation from the contaminated community to a clean community; and I would suggest that the lawsuit was filed in order to be duly compensated for the pains and sufferings they have endured for several years of living in a contaminated landscape. Unfortunately, the entire community was devastated and dispersed by the Katrina disaster in 2005. However, in January 2006, after 13 years of protracted litigation, the judge ruled in favor of the residents. The court held that the ASL site is unreasonably dangerous under Louisiana law. The presiding judge remarked that "the plaintiffs (*Agriculture Street residents*) who are overwhelmingly poor minorities were promised the American dream of first time home ownership. The dream turned into a nightmare." The city of New Orleans, HANO, and HANO's insurers were held liable for paying damages to every member of the ASL class action suit.[6]

Homeowners in Press Park and Gordon Plaza were awarded the full fair pre-Katrina market value of their properties; the court also awarded monetary compensation for diminished property value and emotional distress. The defendants appealed the judgment and the Fourth Circuit Court of Appeal reduced the emotional distress component of the earlier judgment by 50 percent. As of the time of writing this, all homeowners, residents, and other members of the class action suit are directed to complete and submit official claim forms no later than November 13, 2009 to receive their compensation.[7]

The Union Carbide Toxic Disaster in Bhopal, India

Multinational corporations (MNCs) that use, manufacture, and generate toxic chemicals and hazardous wastes tend to seek "the

paths of least resistance," building their plants in the environment
with cheap land, labor, and raw materials, robust tax incentives, and
lax environmental regulation. India, among several other Third
World nations, is particularly an attractive destination for chemical
MNCs. Given the tendency or practice of MNCs to lower safety
and risk management standards in their Third World subsidiar-
ies, a major industrial disaster in these locations seems inevitable.
Chemical waste discharge or toxic waste dumping by MNCs occurs
with impunity on a regular basis in many underdeveloped nations.
Only a megadisaster such as the Bhopal gas poisoning received
extensive media and international spotlight. The Bhopal catastrophe
has been documented extensively in the literature (see Engler, 1985;
Morehouse and Subramaniam, 1986; Rosencranz, 1988; Narayan,
1990; Dhara and Kriebel, 1993; Rao, 1993; Sriramachari, 2004;
Broughton, 2005; Trotter, Day, and Love, 1989; Karan et al., 1986;
Höpfl and Matilal, 2005).

The history of the Union Carbide Corporation (UCC) venture in
India dates back to the early 1930s; the Union Carbide India Limited
(UCIL) was established in 1934 as a diversified manufacturing company, which employed about 9,000 people. The UCIL produced
a range of chemical products including plastic materials, photo-
graphic plates, films, industrial electrodes, batteries, polyester resin,
laminated glass, and machine tools (Morehouse and Subramaniam,
1986: 2; LaPierre and Moro, 2002; Höpfl and Matilal, 2005: 65). It
was licensed to manufacture phosgene, monomethylamine (MMA),
methylisocyanate (MIC), and the pesticide Sevin (Singh and Ghosh,
1987).

In India's quest to maximize the Green Revolution to achieve
self-sufficiency in food production, domestic manufacture of pesticide was attractive to the government who granted UCC permission
to build and operate a pesticide-manufacturing plant in Bhopal, the
capital city of Madhya Pradesh, in 1969. The UCC owned 50.9 percent of the stocks of its India subsidiary, UCIL, and the remaining
49.1 percent was distributed among Indian shareholders including
the government of India, which controlled 22 percent. As indicated
by Morehouse and Subramaniam (1986), foreign investors are generally limited to 40 percent ownership of equity in Indian companies; however, the Indian government waived this requirement for
UCC because of the sophistication and complexity of its technology
and the potential of its products for domestic and export demands.
Initially, this plant was set up to combine and package intermediate
chemicals—MIC and alpha-naphthol diluted with nontoxic powder

to yield a toxic pesticide carbaryl with the trade name Sevin. This facility for manufacturing MIC was sited along the path of least resistance—adjacent to poor communities, "squatter settlements," or shantytowns approximately 2 km from the railway station. More than 80 percent of the residents were considered absolutely poor by official standards and not employed by Union Carbide. As Karan et al. (1986: 202) indicate, untouchables, the lowest of the Hindu castes, and Muslims accounted for a large proportion of the slum dwellers. These people were basically poor, illiterate, unemployed, or employed in menial jobs, and many pursued employment in the informal sector of the urban economy.

Bhopal, India, experienced the worst industrial disaster in history beginning around midnight on December 2 into the early dawn of December 3, 1984, at the Union Carbide's plant. In fact, the tragedy remains unprecedented in the world (Rosencranz, 1988; Trotter et al., 1989; Dhara et al., 2002; Broughton, 2005; Sriramachari, 2004). On that night, a deadly chemical reaction manifested when a substantial amount of water leaked into a storage tank No. E610 containing 42 tons of methyl isocyanate (MIC [CH_3-N=C=O]), which triggered a cascade of uncontrollable exothermic (heat-producing) reactions emitting excessive heat (200 °C or 400 °F) under tremendous pressure (180 psi) and releasing a cloud of poisonous gases into the atmosphere of the plant and the surrounding communities. MIC is a highly volatile, flammable, and toxic chemical compound. A cloud of toxic gases covered more than a 25-square mile area, exposing more than 500,000 residents to toxic gases and killing between 3,800 and 8,000 residents while causing permanent disabilities for more than 150,000 people (Stringer et al., 2002). LaPierre and Moro (2002: 323) described the event of that tragic night as a silent, insidious, and almost discreet massacre; no explosion had shaken the city, no fire had set its sky ablaze; most inhabitants of Bhopal were sleeping peacefully without a siren or any emergency alert concerning the unfolding dread. Thousands of children, older people, and women—the most vulnerable segment of the population—died in their sleep or on the street.

Over the past 25 years, exposure to the poisonous gases and associated toxins have killed over 15,000 people and caused chronic and debilitating morbidity conditions for thousands of men, women, and children in Bhopal. However, it appears the actual number of victims of the Bhopal disaster will remain unknown because the number of people killed, injured, or paralyzed by toxic gases overwhelmed the capacity of existing hospitals, health clinics, and other local, state, and national institutions in India. Morehouse and Subramaniam (1989: 24) describe the dilemma:

> In Hamidia Hospital...there was such a terrible crowd bringing dead bodies that there wasn't even a place to keep them on the floor. As soon as a patient was declared dead, his/her relatives would vanish with the body. At least fifty babies were taken away like this. About 500 to one thousand bodies were taken away before their deaths could be registered...not even half the dead were buried or cremated at official grounds. The government, confronted with a health crisis of catastrophic proportions, had to clear the city of the dead. And that included over four thousand cattle, several hundred dogs, cats, birds, and other animals. Both government and nongovernmental agencies participated in the urgent task of removing corpses from the city. The inevitable consequence of many hands working at this gruesome task is that the government diminished its control over vital statistics.

The number of victims who died in the field, forest, and remote sections of the city remained unaccounted for in any official documentation; and official death records did not account for the number of homeless and transient people killed by the toxic gas.

As noted by Dhara and Dhara (2002: 392), the released MIC cloud probably included heterogeneous toxic decomposition by-products such as carbon monoxide, hydrogen cyanide, nitrogen oxides, and other contaminants including phosgene, MMA, monomethylamine hydrochloride, dimethylamine hydrochloride, trimethylamine hydrochloride, 1,3-dimethyl urea, 1,3-dimethyl isocyanurate, and 1,3, 5-trimethyl isocyanurate (also see Stringer et al., 2002: 6; LaPierre and Moro, 2002: 299). Heavier than air, the MIC constituted the base of the gaseous clouds released by the chemical reactions in tank No. 610. The immediate or acute irritant effects of MIC and other toxic mixtures caused panic, anxiety, disorientation, and confusion, forcing people to run out of their homes into the gas cloud, thus leading to increased exposure to higher doses of the poison.

Without any evacuation plan, emergency warnings, or alarms provided to local residents, more than 3,000 people were estimated to have died instantly in their sleep, mostly the residents of poor slum settlements adjacent to the UCC plant, while thousands woke up with eye and nose irritations and a burning sensation in their lungs as they ran helter-skelter in panic and anxiety seeking escape routes (see figure 6.4). The official death toll was estimated at 6,000 by the government, which most analysts considered to be somewhat conservative (see Broughton, 2005; Dhara and Dhara, 2002; Dhara

Figure 6.4 Victims of Union Carbide India Limited Toxic Gas Explosion in Bhopal, India.

and Kriebel, 1993; Sriramachari, 2004; Miller, 2004). An alternative estimate puts the death toll at 8,000–16,000, judging from the sales of shrouds and cremation wood. In its 2004 report, Amnesty International estimated that between 7000 and 10,000 people died within three days of the disaster, and another 15,000–20,000 have died between 1985 and 2003 due to toxic gas exposure.

About 50,000 to 150,000 people are estimated to have serious chronic, long-term, adverse health problems due to their exposure. The list of health problems associated with MIC exposure among survivors is quite extensive including brain damage, blindness, lung damage, and neurological and psychological problems (Dhara et al., 2002, D'Silva, 2006). As listed in *TED Case Studies*, MIC toxicity led to severe damage to the eyes and lungs, and caused respiratory problems including chronic bronchitis and emphysema. Other serious health problems that have also been reported among the survivors are gastrointestinal problems; neurological disorders such as loss of memory and motor skills; headaches; fatigue; psychiatric problems such as anxiety, depression, irritability, posttraumatic stress disorders, pathological grief reactions, emotional reactions to physical problems, and exacerbation of preexisting psychiatric disorders; musculoskeletal problems; gynecological problems among female victims; increased spontaneous abortion; and increased chromosomal abnormalities (see Morehouse and Subramaniam, 1986; Dhara and Dhara, 2002: 395–398; Broughton, 2005; Kapoor, 1992).[8]

The economic damage from the disaster was estimated at $4.1 billion, and both the Indian government and international analysts viewed the accident as preventable. Rosencranz (1988) points to a diffused "hard" and "soft" jurisprudence of international environmental law, which establishes liability and accountability for environmental hazards. One developing norm of international and domestic law is a strict liability for the injury or harm caused by hazardous industrial activity; this imposes the liability on both state and non-state actors responsible for toxic disaster to pay compensation to the victims. Consistent with other technological disasters, clear identification of the culpable party whose negligence contributed to a catastrophe of this proportion calls for litigation. As can be expected, the toxic cloud still remained in the atmosphere of Bhopal on December 7, 1984, when the first multimillion-dollar lawsuit was filed by American lawyers in a US court. It represented the genesis of several years of legal battles in which the ethical and moral implications of the disaster and its impact on Bhopal's victims would be overlooked. Despite the indisputable fact that the toxic MIC gas, which killed more than 3,800 people and incapacitated and deformed several thousands more, was accidentally released from the UCIL plant to the surrounding communities, the parent company attempted to shift the blame somewhere else to avoid liability. The UCC offered a disgruntled employee sabotage hypothesis, which was vehemently debunked and rejected. Moreover, the company refused to disclose the constituents or chemical composition of the accidentally released gases to the public.

There were several internal problems within the organization that appeared to have precipitated the disaster, including lax enforcement of safety standards, financial struggle, lack of environmental protection, lack of concern for the health of the residents of nearby communities, and substandard staffing of personnel. Davidar (1985: 112) sums up the main reasons for the tragedy as faulty, poorly maintained equipment and inadequately trained personnel, who did not have the knowledge of how to respond to emergency, or even how to run the plant. Staff reduction and budget cut due to financial constraints led to substandard operations within the UCIL Bhopal plant. Thus, as pointed out by Morehouse and Subramaniam (1986: 12), there is compelling evidence showing that both the UCC and its Indian subsidiary in Bhopal were grossly negligent.

The government of India may also be faulted for its lax regulations and inadequate supervision of UCLI operations. Evidence points to the fact that UCC adopted double standards in the operational activities of its US plant relative to UCLI. It has been pointed out that UCC cut resources and employed untested technologies when building the Bhopal plant. Labunska et al. (1999: 6) contend that the responsibility for the accident lies somewhere between the UCC, the Indian operators of UCIL, and the Indian government. Despite the overwhelming evidence indicating culpability of these entities, the clouds of injustice have descended upon the victims of the Bhopal toxic gas tragedy. As mentioned, the first lawsuit was filed against UCC in the US court a few days after the accident, for damages up to $15 billion. Unfortunately, in May 1986, the US Supreme Court ruled that the claims filed on behalf of the victims would not be entertained in the US courts on grounds of *forum non-conveniens*. The term "forum non-conveniens" simply means the discretionary power of a court to refuse jurisdiction if, in the interest of justice and the convenience of the parties, the court considers that the case should proceed in another court, or jurisdiction. In the case of Bhopal, India, since all of the victims, claimants, witnesses, physical evidence, and the impacts of the disaster were all in India, applying this doctrine of forum non-conveniens moved the case to the Indian legal system for trial. This decision has been described as the second tragedy to befall the victims of Bhopal (Trotter et al., 1989: 447; Jasanoff, 2008). Rather than delivering justice to the injured parties, the doctrine of forum non-conveniens has been used effectively by MNCs operating in less developed countries (LDCs) to avoid paying due compensation to the victims of industrial disasters, and thus promote injustice.

As Trotter et al. (1989: 443) indicate, exemplary and punitive damages are rarely awarded in Indian lawsuits and as such, judgments for compensation in Indian courts are typically substantially lower than those in the US courts. Even before a decision was rendered in the suits filed in the US court, in March 1985, the Indian government promulgated the Bhopal Gas Leak Disaster Act, which made the government of India the sole authority to represent the victims in legal proceedings within and outside the country. With this act, the government of India assumed a paternalistic legal role as the guardian of the victims (*parens patriae*) and prevented any class action and independent lawsuits through private lawyers representing the victims. Subsequently, the Indian government filled a $3 billion lawsuit in the Indian court for claims on behalf of the victims. In 1989, the government settled the case with UCC for $470 million without consultation with any of the victims and without UCC admitting guilt or negligence about the accident. UCC also paid $100 million to build a new hospital for the health-care needs of the victims. The $470 million figure was partly based on an erroneous and disputed assertion that only 3,000 people died and about 102,000 afflicted with chronic disabilities (Broughton, 2005). This settlement was affirmed by the Supreme Court of India, which termed it as "just, equitable and reasonable," and settled all claims related to the accident (Höpfl and Matilal, 2005; Trotter et al., 1989). On the contrary, many victims consider this settlement as unfair, grossly inadequate, and a travesty of justice. For one thing, the $470 million settlement was based on inaccurate government estimation of the total number of casualties. Compensation to the victims was handled by the government of India's bureaucratic process. As indicated by Dinham and Sarangi (2002: 90–1), the long process of disbursing compensation focused more on minimizing payouts to victims; and meanwhile, tens of thousands of victims had lost the ability to work, were in huge debt from the costs of medical treatment, and have continuing extensive medical expenses.

Payment of 2,500 to 5,000 US dollars (100,000 to 200,000 Indian rupees) was made to the families of each identifiable dead victim, and victims with injuries were paid between 400 and 800 US dollars (or 25,000 to 35,000 rupees). One survivor of the MIC exposure narrated his personal experience thus:

> I was among the 650,000 persons who filed their claims for compensation. Documentation and medical examination occurred at Bhopal more than once. A bronchial attack on my lungs and on-set of cataract in both eyes were noticed as after-effects of the exposure to MIC. Another half a dozen visits to Bhopal took place for the judicial proceedings. The claim was settled during 1997 with a payment of 800 U.S. dollars (or 35,000 rupees). It virtually meant a negative compensation since much more money had been spent during this period on medication and frequent travels to Bhopal for documentation and judicial proceedings. (Gehlawat, 2005: 3)

The narrator was a professor and chemical engineer, who had a travel layover in Bhopal at about midnight on December 2–3, 1984. The dense toxic cloud from the factory hit the train station after midnight, turning it into a deathtrap for thousands of stranded passengers (LaPierre and Moro, 2002). In light of his high socioeconomic status and his knowledge of Indian government bureaucracy, the negative compensation he experienced may pale in comparison to those of thousands of poor and illiterate claimants who have been marginalized or totally deprived of their due compensation. Many other survivors indicated that the process of compensation claims involved numerous trips to hospitals, government offices, attorneys, banks, and the court. They expressed their frustration as they had to stand and wait for hours in long lines while enduring discrimination, apathy, suspicion, indifference, humiliation, and corruption at the hands of government employees, brokers, middlemen, and attorneys.

THE TOXIC WASTE LEGACY OF BHOPAL DISASTER

December 2–3, 2009, marked the twenty-fifth anniversary of the catastrophe in Bhopal. For a quarter of a century, legal battles persisted in both India and the US courts. The infamous UCIL plant in Bhopal remains a toxic derelict obsolete structure abandoned with thousands of tons of heterogeneous toxic chemicals, poisoning the surrounding soil, underground water supply, and the environment of adjacent communities (Stringer et al., 2002). Although the lease of the factory site has been turned over to the Indian government by UCIL and the company has been acquired by Dow Chemical Company, getting a responsible party to clean up and remediate the old factory site remains problematic (see figure 6.5). After 20 years, the old factory site remains on an overgrown 11-acre forest, conspicuously displaying several corroded tanks and pipes characteristics of a brownfield. Over 425 tons of toxic waste remained buried at the site, posing imminent and continuous danger to all life forms in the vicinity. According to the Greenpeace Research Laboratory's technical report on Bhopal, over one 1,500 tons of toxic chemicals and hazardous industrial waste were dumped or released into the atmosphere of Bhopal between 1969 and 1984 (Stringer et al., 2002; Labunska et al., 1999).

The Indian National Environmental Engineering Research Institute (INEERI) conducted an investigation of the extent of contamination of the site and found the presence and significant concentration of numerous toxic chemical compounds including alpha-naphthol, lindane, Sevin or carbaryl, and termik at several waste disposal sites previously used by UCIL. The institute recommended immediate cleanup of the sites and this recommendation has not been followed. Similar to INEERI, the Greenpeace team conducted an extensive investigation of the site both in 1999 and 2002, reporting an extensive contamination of the site and its surroundings (Labunska et al., 1999; Stringer et al., 2002: 8). Among the contaminants found are

Figure 6.5 Stockpile of Abandoned Toxic Chemicals inside Derelict UCIL Warehouse in Bhopal, India.

Source: Stringer, R., I. Labunska, K. Brigden, and D. Santillo. "Chemical Stockpiles at Union Carbide India Limited in Bhopal: An Investigation." Technical note 12/2002, Greenpeace Research Laboratories, Exeter, UK: Greenpeace. 2002. http://www.greenpeace.to/publications/bhopal%20stockpiles%20report.pdf (November 24, 2009).

metallic and persistent organic compounds in the soil of the site and in the piles of waste left on the factory site. As noted by Stringer et al. (2002: 8), toxic solvents employed in UCIL's production processes have entered underground water over the years, contaminating the wells relied upon by thousands of people in the surrounding communities. Other routes of population exposure to abandoned toxic chemical compounds at the UCLI site encompass inhalation of or skin contact with contaminated dusts, consumption of dairy products from cattle that grazed in and around the site, and direct body contact among children playing in the area. Also, local residents are known to remove soil from the site to fill up the porches and steps outside their houses and therefore may be exposed to contaminants through skin contacts or inhalation of contaminated dusts.

It is noteworthy that the settlement reached by the Indian government and UCC did not include the costs of cleanup and reclamation of the contaminated site, or the relocation of adversely affected communities. The issues of continuous exposure to toxic contamination of the environment around the UCIL site have been raised as additional grounds for seeking damages and environmental remediation measures. As mentioned earlier, the legal battles continue after 20 years of the Bhopal disaster with some surviving victims and grassroots activists and organizations remaining committed to bringing Warren Anderson, the CEO of UCC, and others deemed responsible for Bhopal disaster to justice. Even though environmental awareness and activism have increased in India over the past quarter of a century, the absence of stringent legislative mechanisms in India, such as the Community Right-to-Know Act, the CERCLA known as "Superfund," and other legal measures adopted in the United States following the Love Canal episode and other similar cases, has handicapped the Bhopal residents' ability to force the government to clean up and revitalize the UCIL site and its vicinities.

DIOXIN CONTAMINATION IN SEVESO, ITALY

About eight years prior to the UCIL disaster in Bhopal, a major industrial accident had occurred at the Industrie Chimiche Medionali Societa Azionaria (ICMESA), a Swiss-owned factory located in Meda near Seveso, in Lombardy district in Italy. Givaudan, a Swiss corporation established in 1898, owned ICMESA, and the multinational drug and chemical manufacturer, Hoffman-La Roche, also Swiss-based, was Givaudan's parent company. The latter relied on ICMESA for production of intermediate chemicals such as TCP and aromatic compounds among others used in the manufacture of cosmetics, flavorings, perfumes, and pharmaceutical products. TCP was used as an intermediate chemical in the production of the herbicide trichlorophenoxy acetic acid. It was also a component of hexachlorophene, a bacteriostatic agent produced by Givaudan.

During the summer of 1976, on a hot Saturday July 10 afternoon, an unexpected endothermic (i.e., heat-producing [> 200 ^{0}C or 392 ^{0}F]) reaction in a 2,4,5-trichlorophenol (TCP) reactor took place inside the ICMESA factory. A plume of toxic cloud containing gases mixed with heterogeneous toxic compounds escaped from the plant into the atmosphere of the surrounding communities (Reich, 1991; Signorini et al., 2000). The southeasterly winds blowing at the time pushed the cloud away from Meda into parts of Seveso. In short, the constituents of the cloud fell and dissipated through most of Seveso, affecting vegetation, animals, birds, and an unsuspecting human population who had direct contact with the toxic effluents. While Seveso received the brunt of the toxic cloud, other towns in

Table 6.1   Eleven Towns and Populations Placed under Health Monitoring within and around Contaminated Zones

Towns	Zone A	Zone B	Zone R	Outside (Reference)	Total
Seveso	681	626	7945	7738	16990
Meda	55	***	4017	15493	19565
Cesano Maderno	***	731	14991	16111	33833
Desio	***	1342	4608	27061	33011
Barlassina	***	***	72	5559	5631
Bovisio	***	***	167	11058	11225
Lentate	***	***	***	13037	13037
Muggio	***	***	***	18690	18690
Nova Milanese	***	***	***	18467	18467
Seregno	***	***	***	36838	36838
Varedo	***	***	***	11841	11841

Source: Adapted from Homberger, E., G. Reggiano, J. Sambeth, and H. K. Wipf. "The Seveso Accident: Its Nature, Extent and Consequences." *Annals of Occupational Hygiene* 22, no. 4 (1979): 333.

the vicinity such as Meda, Desio, Cesano Maderno, Barlassina, and Bovisio Masciago were also not spared. As illustrated in table 6.1, 11 communities along the countryside between Milan and Lake Como were directly or indirectly involved in the toxic contamination and its aftermath. The population of the townships and contaminated and noncontaminated communities and zones are illustrated in the table.

The ICMESA cloud of toxic chemicals descended upon a densely populated area like a dense fog, and caused acute effects including coughing, irritation of the eyes, headaches, dizziness, and diarrhea. Many children, especially those who were playing outdoors during the afternoon of the explosion, developed a severe case of chloracne—an acne-like skin disfiguration in which the skin develops blackheads, papules, pustules, and other blemishes (Hernan, 2010: 52; Pesatori et al., 2009; Consonni et al., 2008).

As it turned out, the constituents of the Saturday afternoon cloud were later identified as 2,3,7,8-tetrachloro-dibenzo-p-dioxin (TCDD) or simply "dioxin," sodium hydroxide, ethylene glycol, and sodium trichlorophenate. It was estimated that more than 34 kg of dioxin was released into the environment of Seveso, contaminating everything in its path (Bertazzi et al., 1998). Dioxin is considered one of the most dangerous man-made chemical compounds with numerous adverse health effects on human and nonhuman species. As suggested by

DeMarchi et al. (1996), the image of dioxin was in many ways similar to that of radioactivity—characterized by its invisibility, poisoning at the microscopic level, and it has been implicated as Agent Orange in the Vietnam War. Givaudan personnel knew about the release of dioxin in a number of previous accidents at other TCP plants in Austria, England, Germany, the Netherlands, and the United States from the 1940s through the mid-1970s, with horrible health consequences for the exposed victims. However, this knowledge was not shared immediately with the public officials. In addition to chloracne, there were reports of deaths, cancers, liver damage, birth defects, emphysema, myocardial degeneration, renal damage, chromosome breaks, immune system dysfunction, hypertension, memory loss, and acute poisoning among the population exposed to dioxin (Reich, 1991: 104; Homberger et al., 1979: 332; Signorini et al., 2000).

More than 37,000 people in Seveso area were exposed to this highly toxic substance. A large number of people were indulging in outdoor activities, and families had opened their doors and windows to let in the fresh breeze on the day this contamination episode took place. Many children were playing outside in the yard when the toxic cloud engulfed the area. Therefore, a large population across age and gender was seriously exposed to the toxic cloud that descended upon Seveso and consequently presented dioxin-induced adverse health problems. As indicated by Signorini et al. (2000: 263), those who happened to be in the path of the cloud developed acute symptoms including nausea, headache, eye irritation, and within a few days, children developed severe cases of skin disorders—skin lesions, chloracne, and rashes. Free-range animals died, and plants wilted upon contact with the toxic effluents. Hernan (2010: 50) describes the effects of acute toxicity of the cloud in Seveso on living things, including humans, as birds were seen dropping dead from the sky, pets walked drunkenly as if disoriented, and numerous domestic animals died within 24 hours of exposure. Rabbits spilled blood from their mouths and rectums, while plants turned brown, as if burned by fire. Children developed sores all over their bodies, with several hospitalized, and numerous adults presented cases of nausea, vomiting, and liver and kidney problems.

There was no emergency preparedness, and people of Seveso were not immediately informed about the nature or extent of contamination that descended upon their community, families, homes, and gardens. The mayors of Seveso and Meda were briefed about the accident by the ICMESA's production managers on Sunday, July 11, and they explained the event in general terms that some "aerosol mixture" and a chemical known as "trichlorophenol" had leaked from the factory to the surrounding neighborhoods. Business resumed as usual the following Monday at the factory without the workers being dismissed or asked to stay away from the contaminated plant. The only precaution taken was to block the entrance to Department B, which produced the TCP. A new fence and signs reading "No Entrance," were placed around the facility. The factory continued to operate through the week until Friday when more than 60 percent of workers declared their intention to strike until the management provided information about the degree of danger or risk faced by workers at the plant due to Saturday's explosion. The strike shut down the factory and a couple of days later, the mayor of Meda issued an ordinance to close down the factory.

As noted by Reich (1991: 100), a letter was written and dispatched to the health officials serving both Meda and Seveso by ICMESA's management stating that the cause of Saturday's accident could not be determined and it was only known to have resulted from an inadvertent exothermic chemical reaction in a reactor that was set at a cooling phase, consistent with the normal operating procedure. For several days after the accident, there was no public announcement or direct communication concerning the magnitude of risk faced by the residents of Seveso and Meda. Some people who were concerned enough to go to the ICMESA factory to learn about what had occurred were told that there was nothing to worry about. Yet, people living in close proximity to the factory were advised to avoid eating their homegrown produce. Clearly, ICMESA officials did not demonstrate high corporate social responsibility toward the people of Seveso, as they placed protecting their corporate image over protecting public health.

Several days passed before the company collected samples of materials in the heavily contaminated zones and sent them to the Givaudan research company's analytical laboratory in Switzerland for analysis. On Tuesday, July 20, Givaudan disclosed the result of the analysis conducted indicating the presence of dioxin at high levels both within the ICMESA facility and beyond the factory. The next day, after a conference with the health commissioner for the Lombardy region, the mayor of Seveso passed an order to close several streets in the contaminated area to all traffic. The authorities moved promptly to respond to the crisis gradually unfolding. Both the ICMESA's factory director and production manager were arrested and booked on criminal charges of being responsible for causing a disaster as a result of their culpable negligence (*disastro colposo*) (Reich, 1991: 103;

Hernan, 2010: 50). Based on the report of the analytical results by the Givaudan research laboratory in Switzerland, the health commissioner issued a press release on plans for addressing the contamination— keeping ICMESA plant closed, mapping the entire area contaminated by dioxin, proceeding with health-protection measures for the exposed population, and implementing measures to contain and control polluted agricultural produce and dairy products (Homberger et al., 1979).

With the results of the sample analysis revealing varying degrees of dioxin toxicity, the authorities moved to divide the affected areas into four zones. The area immediately south of the ICMESA factory with the highest dioxin toxicity (i.e., > 50 μg/m^2) was designated as zone A, wherein the residents were ordered to evacuate and leave all their belongings, food, pets, and livestock behind (Bertazzi et al., 1998; Baccarelli et al., 2005; Pesatori et al., 2009). Italian soldiers installed barbed-wire barriers around zone A, and patrolled the area to keep people out of the contaminated zone. The residents (212 families and 736 individuals) were evacuated from zone A between July 26 and August 2 and sheltered in hotels outside Milan. Some houses were demolished in this zone during the period (Signorini et al., 2000). A less contaminated area (with TCDD soil concentration between 5 and 50 μg/m^2) to the south of ICMESA was labeled zone B. Thus, the borderline separating zones A and B was set at an average of 50 μg of TCDD per square meter of the soil (Reich, 1991; Homberger et al., 1979).

The residents of zone B were advised to follow certain health precautions. Children and pregnant women in this zone were ordered to be removed from the area during the day, and they could return at night to minimize their exposure to dioxin; they were also advised not to eat any produce or animals raised in the area. The area with negligible or insignificant soil concentration of TCDD, which comprised several thousand acres of land and about 31,800 inhabitants, was designated zone R, to be monitored without restriction of people's movement. The areas farther away from ICMESA factory, deemed noncontaminated, were the reference zone used by numerous postdisaster studies of the effects of TCDD on human and animal health. As is typical of an industrial disaster of this magnitude, there were conflicts at every level; and anger, frustration, stress, anxiety, and fear of short-term and long-term unknown health effects of TCDD exposure were expressed by the population of the entire area.

The stigma of a contaminated community became apparent when the neighboring uncontaminated towns and cities began to view all the inhabitants of Seveso as carriers of dioxin. Consequently, all goods or merchandise from the entire Seveso area, whether contaminated or uncontaminated, were rejected due to fear of them having been exposed to dioxin contamination. People avoided traveling to the area. Thus, because people were perceived as carriers of dioxin, it took on the appearance of a dreadful disease—a plague to be avoided at all costs. Children with scarred faces due to chloracne experienced significant social distance from others, and those without any visible symptoms suffered discrimination at hotels, restaurants, border crossings, and in other social interactions throughout Italy. Thus, both the visible and invisible toxic tainting produced social discrimination, unleashing a secondary and tertiary social victimization consequent upon a primary toxic victimization (Lifton, 1976: 17; Reich, 1991: 171).

Even though there was no human death associated with the accident and the acute effects of the toxic cloud, several thousand small animals were casualties, including more than 2,000 rabbits, other livestock, and pets. Animals that did not die of acute poisoning were slaughtered by the authorities. More than 50,000 animals were exterminated to prevent the entry of TCDD into the food chain (Homberger et al., 1979).

Women and married couples in contaminated area were advised to avoid pregnancies, and those already pregnant were asked to report to a prenatal clinic for preventive medicine and to undergo strict observation. Those women already pregnant were given the option of considering therapeutic abortion to avoid a dreaded congenital malformation of their newborns, which may occur as a result of dioxin toxicity. As noted by Hernan (2010: 54) and Reich (1991), about 34 pregnant women were granted permission to have therapeutic abortions and several hundred probably had abortions without permissions either in Italy or outside the country. Thus, the actual number of abortions consequent upon the TCDD contamination in Seveso may never be known.

Unlike Bhopal, India, justice moved very swiftly to resolve the legal aspects of the Seveso accident. As mentioned earlier, the company's executives—the managing director, chairman, and three personnel of ICMESA and its parent company—were charged with criminal negligence and convicted by an Italian court, with a prison sentence ranging from two years and six months to five years. However, the prison terms were suspended in the subsequent appeal. And, in an illegal "act of retaliation," ICMESA's director of production, Paolo Paoletti, was murdered by an Italian terrorist group in February 1980. Reich (1991: 136–137) summarized the legal settlement thus:

> In the 1980s, Givaudan paid settlements to various parties involved in the ICMESA disaster. The towns of Meda, Cesano, Maderno, and Desio received payments to compensate for administrative costs. Seveso, in hope of reaching a quick settlement, filed its case in Switzerland. Most private individuals negotiated with Givaudan for out-of-court settlements for property damage. At the end of 1982, the company had paid $6.1 million for damage to agriculture, $3.5 million for damage to industries in zones A & B, and $8.9 million for damage to individuals including land, houses, furniture, decontamination, medical and other related expenses.

Following the legal settlement, decontamination and cleanup of contaminated zones moved swiftly relative to any other technological disasters discussed in this book. The unit where TCDD or dioxin escaped was dismantled. In late August 1982, workers in protective airtight suits were sent to dismantle the reactor and secure the TCDD-contaminated material in lead drums. The most toxic component of the material from the reactor (about 2.2 metric tons) was loaded into 41 drums and sealed for disposal outside the area. In a surreptitious fashion similar to what occurred in Koko, Nigeria, discussed in the next segment, the 41 drums of dioxin were taken out of Italy on September 10, and they disappeared in Europe for some time, until they were found as illegal dumping in a slaughterhouse in a countryside in France by the French authorities.

The Case of Toxic Waste Dumping at Koko, Nigeria

The case of a small fishing village of Koko, Nigeria, gained international attention in 1988 when it became exposed as a depot for highly toxic waste generated and exported by two Italian firms (Adeola, 2001). This toxic waste episode illustrates that poverty is often a critical factor enticing people into accepting hazardous waste for cash. In 1987, two Italian MNCs—Ecomar and Jelly Wax—enticed a Nigerian businessman, Sunday Nana, into an agreement to use his residential property located in Koko, Nigeria, for the storage of about 18,000 drums of hazardous wastes disguised as building materials and allied chemicals for $100 a month (Greenpeace, 1994). This offer was too attractive to be rejected, especially in a country with a per capita gross national product (GNP) of less than $300 at the time.

Upon media exposure of this case, the Nigerian authorities later discovered that the illegal toxic wastes included a wide range of lethal substances including PCBs, dioxin, methyl melamine, dimethyl formaldehyde, and asbestos fibers (Greenpeace, 1993; Frey, 1994–95; Ihonvbere, 1994–95). Most of the drums were damaged and were leaking toxic substances into the soil (Lipman, 2002). Over 100 employees of the Nigerian Port Authority were deployed to remove the wastes, which had been deposited for several months prior to the discovery, thus contaminating the soil, water, and air in the vicinity of the dump. Even though the government provided the workers with protective devices, the wastes were more toxic than many had suspected. The exposure suffered by the workers and the residents of Koko, including the host, led to severe adverse health consequences including burns, nausea, paralysis, premature births, and deaths, including the death of Sunday Nana, the host, who died of cancer of the throat. Other health hazards associated with the dump include birth defects, brain damage, cancer, stunted growth, and other pathological conditions.

Through international politics and diplomacy, these wastes were removed and sent back to Italy at the expense of the Italian government. In July of 1988, two ships, the Karin B and the Deepsea Carrier, loaded the toxic wastes and began the process of transporting them from Nigeria back to Italy. However, for the residents of Koko, the damage had already been done; land within a 500 meter radius of the dump site was officially declared unsafe and unfit for use. In addition to adverse health consequences, Koko became stigmatized as a toxic place, to be avoided. The vernacular of reference to this town included any combination of the terms: "toxic," "hazardous," "sick," "poisonous," "radioactive," "dreadful," "corrosive," and "dangerous." As noted by Ihonvbere (1994–95: 211):

> The public started avoiding Koko town. Commercial vehicles would not stop at the road junction or intersection leading to the town, and private car owners would hold their breath and wind up their windows as they approach the town. Traders stayed away from the community market and visitors to Koko were avoided like plague. *The only bank in the town closed its offices, and non-indigenes fled the town.* (emphasis in italics)

Thus, there was anxiety and feelings of destitution, isolation, anger, and rejection among the local residents. The adverse social and psychological impacts of Ecomar and Jelly Wax's dumping on Koko still linger even years after the incident. Although the role of Italian MNCs as the culprit seems glaring, the underlying problems, especially in Nigeria, at that time included government corruption, bribery, inefficiency, and abuse of power by military and public officials at the expense of poor innocent people.

The role of the media was pivotal in galvanizing public response to Koko's contamination episode. Among the responses, public rallies and protests were launched demanding immediate evacuation of the toxic wastes and free medical screening and treatment of the potential victims, while rejecting any idea to evacuate or relocate the population of Koko. Several grassroots antitoxic groups developed including the Koko Defense Group and People United to Save Koko, among others, organized to promote environmental justice in the region. In this case, the community contamination episode reflects mostly environmental injustice (undue imposition of toxic wastes and associated adverse health effects on innocent people). This episode of transnational dumping of toxic waste spurred the creation of the Basel Convention to regulate and curb transnational movement of hazardous waste. Unfortunately, not all countries in the world have ratified the convention.

Similar to other cases discussed in this chapter, litigation ensued in Koko several decades after the contamination episode. A class action lawsuit was filed against the Nigeria Port Authority by some of the victims. After 21 years of legal maneuver, a settlement of 39.7 million naira or $264,666 was reached to compensate 94 victims of the 1988 toxic dumping in Koko (*This Day*, 2008; *Vanguard*, 2008).[9] The compensation was not distributed equitably among the recipients who were either members of the staff or officers of the Nigerian Port Authority. Many local residents of Koko who were exposed to the toxic waste or whose property became contaminated, and who suffered associated adverse health effects, were excluded. Many local victims of this contamination episode have been forgotten by the authority.

Maquiladoras along the US-Mexico Border

By definition, *maquiladoras* are factories in Mexico where imported component parts are assembled and packaged for exports.[10] They are fully owned by foreign MNCs mostly based in the United States, Japan, and Western Europe. The maquiladoras emerged under the Border Industrialization Program (BIP) introduced in 1965 by the

government of Mexico with the aim of promoting industrial development, technology transfers, employment opportunities, and a better
standard of living for people in the northern part of the country.
Foreign firms are required to register with the government of Mexico
to be considered maquiladoras (GAO, 2003). Once registered, MNCs
were given considerable investment incentives to set up facilities
within 28 km (17 miles) of the US-Mexico border—an area stretch-
ing nearly 2,100 miles (3, 379 km) from the Pacific Coast to the
Gulf of Mexico across California, Arizona, New Mexico, and Texas in
the United States and Baja California, Sonora, Chihuahua, Coahuila,
Nuevo Leon, and Tamaulipas, in Mexico (see the map in figure
6.6). The BIP permits these companies to import component parts,
machinery, raw materials, and unfinished materials duty-free for final
manufacturing and assembly for export, provided the wastes generated are returned to the country of origin. Other incentives included
tax breaks, cheap lands, low-wage labor pool, and lax environmental
regulations. Within 10 years of launching BIP, Mexico experienced
a rapid (16 percent) growth in maquiladoras relative to a 2 percent
growth on the US side of the border (Williams and Homedes, 2001:
321). Among the products of these factories are automobile parts,
automotive batteries, cardboard boxes, circuit boards, compact discs
(CDs), computer keyboards, copy machines, children's toys, dispos-
able lighters, eye glasses, fiber optic cables, picture frames, pharmaceutical and medical supplies, pool and jacuzzi covers, polystyrene
foam packaging, telephone, television and other electronic goods,
thermometers, textiles and apparel, vacuum cleaners, speakers, and
much more, mostly destined for US consumers.[11]

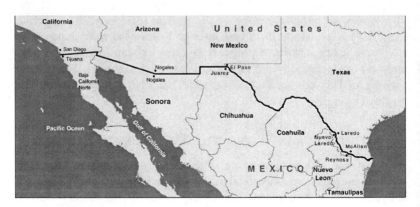

Figure 6.6   The Map of US-Mexico Border Regions of *Maquiladoras*.
Source: GAO.

The 1994 North American Free Trade Agreement (NAFTA) between the United States, Canada, and Mexico was a landmark economic integration between countries in the global North and global South. This agreement became a catalyst boosting the proliferation of maquiladoras in northern border towns and cities of Mexico including Tijuana, Ciudad Juarez, Mexicali, Matamoros, and Reynosa among others. Presently, there are more than 3,500 maquiladoras operating in Mexico's northern border zone. After decades of its implementation, most of the promises of NAFTA, such as better paying jobs, lower prices for consumer goods, economic security, technology transfer, better public health and environmental quality, and decreased emigration among Mexican citizens, especially to the United States, have not been kept. On the contrary, economic instability, dependency, abject poverty, human rights abuse, deplorable working conditions, social and environmental injustices, toxic waste releases, exposures of people to toxic effluents, adverse health conditions, and contamination of flora and fauna, rivers, and underground water supplies have been the latent pathological consequences of maquiladoras and NAFTA for most communities on the Mexico side of the US-Mexico border. The toxic chemicals used in maquiladora industry encompass heavy metals such as lead, cadmium, mercury, nickel, and solvents such as trichloroethylene, 1,1,1-trichloroethane, acids, paints, oils, resins, and plastics, which pose severe risks to humans and other species upon contact.

As Frey (2003: 330) notes, occupational and environmental exposure to the hazards generated by the industry and the consequent health dysfunctions are yet to be fully determined. What has been established in the literature, however, is that populations exposed to toxic industrial effluents are at higher risk of death, injury, and morbidity including various types of site-specific cancers, dermatological disorders, breathing problems, neurobehavioral disorders, mental retardation, especially among children exposed to lead, reproductive abnormalities including birth defects (such as spina bifida and anencephaly), spontaneous abortions, miscarriages, organ dysfunctions, genetic aberrations, and immune system disorders. The plight of the communities hosting maquiladora facilities in northern Mexico have attracted scholarly attention with several researchers including Carruthers (2008), Grineski and Collins (2008), and Grineski, et al. (2010) focusing on the patterns of social and environmental injustices and environmental justice activism, while others such as Frey (2003) and Williams

and Homedes (2001) addressed the transfer of hazardous industries to export processing zones and the exposure of host communities to myriads of toxic effluents and the consequent health
and environmental problems. Prior to 1994, all hazardous wastes
generated by maquiladoras were required to be exported back to
the country of origin as previously mentioned. However, with the
regime of NAFTA starting in 1994 to the present, all wastes generated by these factories are managed and contained within Mexico.
Among the approximately 3,500 foreign-owned processing plants
in Northern Mexico townships, one particular case stands out
in terms of what can go wrong with such a scheme. The case of
Metales y. Derivados has been narrated in the literature (see EHC,
2004; Simpson, 2007; Carruthers, 2008). This case remains particularly germane to our understanding of community contamination and the collaborated grassroots environmental justice struggle
along the US-Mexico border towns.

THE CASE OF METALES Y. DERIVADOS

The history of Metales y. Derivados has been chronicled along
with the varying achievements and weaknesses of environmental
justice struggles over its activities and environmental quality viola-
tions around the industrial park on the Mesa de Otay, adjacent to
Colonial Chilpancingo.[12] This company was incorporated in 1972
as a maquiladora involved in lead and copper smelting processes in
Tijuana, Baja California, Mexico. It was owned by New Frontier
Trading Corporation, a US wholesale metal business based in San
Diego, California. The primary operations of Metales y Derivados
consisted of importing and recovering lead, copper, and phospho-
rous through the smelting of used lead acid batteries, lead oxide,
and other scrap materials to produce refined lead and phosphorized
copper granules for export.[13] In 1986, the factory was moved from
the Tijuana neighborhood of La Gloria to the Ciudad Industrial
Nueva Tijuana industrial park on Mesa de Otay approximately
150–600 yards from Colonial Chilpancingo's 10,000 low-income
residents. As mentioned by Carruthers (2008: 557) and Simpson
(2007: 162), Metales y Derivados stands out as one of the US-
Mexico border's most conspicuous symbols of a maquiladora threat
to communities' environment and public health. This company
befouled the environment of its host communities and exposed
thousands of workers and residents to myriads of acute and chronic
health risks.

For more than 20 years, residents of host communities to Metales y Derivados expressed concerns and launched complaints to the local and national officials about the potential threats to human health and the environment posed by this company's activities. A study reported in the *Washington Post* found lead levels 3,000 times higher than US thresholds, and cadmium 1,000 times higher than acceptable standards (see Sullivan, 2003). Complaints by local residents and environmental justice activists ranged from acute lead poisoning and elevated blood levels among children living near the plants to unsafe and hazardous industrial practices, which prompted Mexico's environmental protection agency or the Federal Ministry for Environmental Protection (PROFEPA) to launch an investigation and a series of inspections of the factory. Among the numerous violations of hazardous waste management obligations of this company documented by PROFEPA are failure to obtain registration as a hazardous waste generator, lack of emission control systems, inadequate storage and disposal of hazardous wastes, failure to control runoff of acids with high concentration of lead salts, failure to report hazardous waste spills, discharges, and infiltration, and blatant violation of acceptable cadmium and lead limits under the Mexican law. Other serious infractions included lack of operating license, failure to maintain log of hazardous waste generation and movement, illegal toxic waste dumping, and lack of trained certified personnel responsible for hazardous waste management. A series of sanctions were imposed by PROFEPA, which the company failed to fully comply with. On March 28, 1994, a complete shutdown of Metales y Derivados plant was ordered by PROFEPA, and an arrest warrant was issued against the owner for alleged violations of Mexico's environmental laws. The owner absconded to the United States to avoid prosecution, leaving behind approximately 24,000 tons of a mixture of hazardous wastes including over 7,000 tons of lead slag (EHC, 2004, Simpson, 2007; Caruthers, 2008). The abandoned plant continued to leach toxic heavy metals and other hazardous compounds into the soils and water of the surrounding communities.

In 1998, two community groups (San Diego's EHC and Colonial Chilpancingo's Padre Canyon Restoration Committee) applied to the North American Commission for Environmental Cooperation (NACEC), the principal institution of NAFTA's environmental "side agreement," for help in cleaning up the contaminated site.[14] The binational border community groups claimed that the parent company, the New Frontier Trading Corporation, and its subsidiary,

Metales y Derivados, have failed to repatriate to the United States the toxic waste they generated, as required under the Mexican law and the 1983 US-Mexico La Paz Agreement that required all waste generated by maquiladoras to be returned to the country of origin.

Four years passed before the NACEC released its report, which confirmed the community groups' suspicion that the abandoned site presented serious risks to humans and the environment. Most importantly, it recommended cleanup and remediation of the site. Through mutual agreement and cooperation (involving the Collective Chilpancingo Pro Justicia Ambiental/or the Chilpancingo Environmental Justice Collective), the Mexican PROFEPA, and US EPA, a comprehensive cleanup of the contaminated site was launched in 2004. The remediation was expected to be completed within five years, with Mexico's state and federal governments contributing $500,000 and the US EPA agreeing to contribute $85,000 toward remediation estimated to cost between 5 and 10 million US dollars. The state of Baja California appropriated the contaminated site and assumed responsibility for restoration. As reported by the Environmental News Service (ENS), in January 2009, both the US EPA and Mexico's environmental ministry, jointly marked the cleanup of Metales y Derivados's abandoned contaminated site. While the host communities and grassroots groups demanded that all hazardous wastes from the site be repatriated to the United States, the majority was treated and capped onsite, consistent with the agreed-upon remediation plan. As mentioned by Carruthers (2008), the Metales y Derivados case reflects a much broader array of problems experienced by communities, organizations, and individuals involved in critical environmental justice struggles throughout the entire US-Mexico border enclave.

CHAPTER SUMMARY

Both in the industrialized and the underdeveloped countries around the globe, an increasing frequency and magnitude of catastrophic accidents, human miscalculations, and gross negligence have led to the contamination of communities where people live, work, learn, and play. Numerous communities befouled and degraded by industrial disasters or stigmatized by the mere revelation of being built upon toxic waste dumps have caused untold hardship, including loss of life, property, and acute and chronic physical and mental health dysfunctions among the victims. The twentieth century and the early twenty-first century in particular have witnessed too many disasters of technological/anthropogenic and natural etiologies around the

globe. This chapter is mostly focused on the former while acknowledging the frequency and devastation of the latter on social systems.

In this chapter, a number of selected cases of industrial disasters
and toxic contamination of communities around the world have been
presented. The cases selected are by no means exhaustive as page limitation makes it impractical to cover all the cases of industrial disasters
and toxic contamination during these two centuries. Nevertheless,
the specific cases included have been covered extensively in the mass
media, attracted scholarly interest, and are considered prominent
cases with important lessons. One common thread among the cases
is the anthropogenic root of each catastrophe revealing the latent dysfunctions of modern science and technology. Human error, negligence, or carelessness is often to blame for technological failures and
the consequent surprises—as found at the Love Canal, New York,
Bhopal, India, Seveso, Italy, and Minamata, Japan. Clearly, the disposal of hazardous waste has become a major problem for countries
that are large waste producers, such as the United States. Another
well-documented episode of hazardous waste disposal problem is the
"Valley of the Drums" in Kentucky, where 17,000 drums of hazardous waste was dumped into a seven-acre site, which contaminated
the soil and water of the area (Lipman, 2002). The cross-cultural,
comparative approach to community toxic contamination reveals that
more affluent societies tend to respond quickly and recover much
faster from either industrial accidents or community contamination
episodes relative to their counterparts in less developed nations where
due compensation and reclamation of afflicted community may be
impossible. The community-specific and event-specific nature of each
of the cases presented in this chapter makes any generalization beyond
the specific episode treacherous.

7

THE REGULATORY FRAMEWORKS

The systems of environmental legislation specifically addressing hazardous and toxic waste problems are a recent development in the United States. Most of the legislative responses to toxic substances and hazardous wastes occurred in the post–World War II era, especially from the mid-1960s to the present. A series of environmental and technological disasters of epic proportions—including Bhopal, India, Love Canal, New York, Seveso, Italy, Chernobyl, Soviet Union, Exxon Valdez, Alaska, and numerous others, causing massive ecological destruction in both the built and natural environments galvanized public outcry and support for governmental regulation of environmental hazards both in the United States and across the globe. In the United States, environmental laws and their provisions are initiated, implemented, and enforced at the federal level. The Environmental Protection Agency (EPA) is the arm of the federal government charged with the task of enforcing environmental laws and regulations. However, state and local environmental protection agencies often act as proxies in the implementation and enforcement process. In fact, some states enact their own environmental legislation to protect unique environments under their jurisdiction. In general, however, state laws and regulation are patterned after the federal statutes. Thus, both the federal and the state government environmental agencies are charged with the tasks of implementing and enforcing applicable environmental laws.

Within the European Union (EU), environmental laws and their provisions are developed and adopted by member states. The international laws regulating toxic waste and industrial disasters are recent developments and most of these are in the form of agreements, which often lack the status of *jus cogens*. The primary aim of this chapter is to briefly sketch the key legislative instruments regulating toxic waste and other environmental hazards in the United States and internationally. Specifically, the key US statutes on hazardous wastes, the

Rotterdam Convention, and the Basel Convention are briefly summarized. The EU directives and international regulation of toxic substances and transboundary movements of environmental hazards are also examined. Lastly, the Helsinki Convention on the transboundary effects of industrial accident is addressed.

LEGISLATIVE RESPONSES TO TOXIC WASTE PROBLEMS IN UNITED STATES

From 1965 to the present, several laws and legislative amendments specifically addressing toxic and hazardous waste releases, ecological disruptions, human habitat destruction, and numerous environmentally induced health risks across the United States were established. Prominent among the environmental statutes designed to regulate hazardous and toxic substances in the environment are the Toxic Substances Control Act (TSCA), Clean Air Act, Clean Water Act, Resource Conservation and Recovery Act (RCRA), the Comprehensive Emergency Response, Compensation and Liability Act (CERCLA), and the Pollution Prevention Act. Each of these statutes is briefly discussed in the following section. A brief summary of selected federal statutes regulating hazardous/toxic substances is presented in table 7.1.

THE TOXIC SUBSTANCES CONTROL ACT (TSCA)

The TSCA of 1976 establishes the toxic substances program to be administered by the EPA. It was primarily aimed at the chemical industry, to regulate the manufacture, importation, use, and disposal of chemical substances. The EPA is authorized to set up regulations pertaining to the testing of chemicals and mixtures of chemicals, premanufacture notification for new chemical substances or significant new applications of existing substances, control of chemicals that pose an imminent hazard, and record keeping and reporting requirements. Furthermore, the act establishes a committee to develop a priority list of chemical substances to be tested. Priority is assigned to substances known to be carcinogenic, mutagenic, or teratogenic. The committee can list up to 50 chemicals, which must be tested within a year. Chemical manufacturers are required to notify the EPA at least 90 days prior to manufacturing or importing a new chemical substance. The EPA has the authority to prohibit the manufacture, sale, use, or disposal of new chemical substances if there is evidence that they pose an unreasonable risk to health or the environment. TSCA

Table 7.1 Selected Federal Statutes Regulating Hazardous and Toxic Substances in the United States, 1965–2001

Statutes	Year Passed	Agency	Enforcement Target of Regulation
Solid Waste Disposal Act	1965	EPA	Municipal wastes
Clean Air Act	1965	EPA	Air pollution; air emission from area, stationary and mobiles sources
	1970		
	1977		
Clean Water Act	1972	EPA	Water pollution; restoration of water quality
Water Pollution Control Act	1977	EPA	
Hazardous Materials Transportation Act	1975	DOT	Interstate transportation of hazardous materials in commerce
	1994		
Toxic Substances Control Act (TSCA)	1976	EPA	Existing and impending new toxic chemical hazards, hazardous substances including asbestos, radon, lead, PCBs, etc.
Resource Conservation and Recovery Art (RCRA)	1976	EPA	Cradle to grave management of wastes and recovery of resources
Comprehensive Environmental Response, Compensation, and Liability Act (CERCLA)	1980	EPA	Hazardous waste sites; compensation for and restoration of contaminated sites
Hazardous and Solid Waste Amendments (HSWA)	1984	EPA	Underground storage tank and waste treatment, storage, and disposal facilities
Asbestos Hazard Emergency Response Act (AHERA)	1986	EPA	Asbestos abatement in schools and commercial buildings
Lead Exposure Reduction Act (LERA)	1992	EPA	Reduction of lead contamination and toxicity.
Emergency Planning and Community Right-to-Know Act (EPCRA)	1986	EPA	Emergency preparedness, minimization of chemical accidents, and dissemination of information to the community
Pollution Prevention Act	1986	EPA	Waste reduction, pollution prevention education, training, and information exchange

Source: Adapted from various sources including Plater, Z. J. B., R. H. Abrams, and W. Goldfarb. *Environmental Law and Policy: A Coursebook on Nature, Law, and Society*. St. Paul, MN: West Publishing Company, 1992; Jain, R. K., L. V. Urban, and H. E, Balbach. *Environmental Assessment*. New York, New York: McGraw-Hill, 1993; Kubasek, Nancy K. and Gary S. Silverman. *Environmental Law*. Upper Saddle, NJ: Prentice Hall, 2000; Chapman, Stephen R. *Environmental Law and Policy*. Upper Saddle River, NJ: Prentice Hall, 1998; and http://tis.eh.doe.gov/oepa/law_sum/TSCA.HTM

also regulates the labeling and disposal of polychlorinated biphenyls (PCBs) and prohibits their production and distribution since July 1979 (Jain et al., 1993; Plater et al., 1992).

Three major amendments to TSCA include the Asbestos Hazard Emergency Response Act (AHERA) of 1986, Indoor Radon Abatement Act of 1988 (P.L. 100–551), and Lead Exposure Reduction Act (LERA) of 1992 (P.L. 102–550). The AHERA of 1986 authorizes the EPA to amend the TSCA regulations to impose more stringent requirements on asbestos abatement in schools. It provides for mandatory periodic inspection of asbestos in schools and implementation of appropriate response actions. Furthermore, it requires the EPA to determine the extent of risk to human health posed by asbestos in public and commercial buildings and the means to respond to such risks. AHERA was amended in 1990 by the Asbestos School Hazard Abatement Reauthorization Act (ASHARA) to require accreditation of individuals who inspect for asbestos-containing materials in schools, public, and commercial buildings. The Indoor Radon Abatement Act was added to the TSCA in 1988. The purpose of this amendment was to assist states in responding to the threat to human health by exposure to radon. The EPA is required to publish an updated citizen's guide to the health risk posed by radon, and to conduct studies of the radon levels in public buildings including schools and federal buildings. The LERA (P.L. 102–550) was added to TSCA in 1992 to reduce environmental lead contamination and prevent health problems associated with lead toxicity due to exposure, especially among children. The provisions of this law include identification of lead-based paint hazards, definition or specification of levels of lead allowed in various products, and establishment of state programs for the monitoring and abatement of lead-exposure levels (including training and certification of lead abatement personnel (see table 7.1).[1]

THE RESOURCE CONSERVATION AND RECOVERY ACT (RCRA)

The RCRA was a landmark legislation enacted in 1976 to establish the federal program regulating solid and hazardous waste management from the point of generation to the disposal sites, or from the cradle to the grave. RCRA actually amended the Solid Waste Disposal Act (SWDA) of 1965 designed to address solid waste disposal problems in the United States. The Resource Recovery Act of 1970 amended SWDA prior to the passage of the RCRA. This legislation has been amended in 1984 and 1992 by the Congress to expand the scope and requirements of the law and in 1996 by the Land Disposal Program Flexibility Act (LDPFA). Thus, RCRA is a complex legislation encompassing the first federal solid waste law and all the subsequent amendments. In its original formulation, RCRA was designed to control hazardous wastes from the cradle to the grave and to provide for resources recovery from the environment (see Jain et al., 1993; Chapman, 1998). It defines solid and hazardous waste, authorizes EPA to set standards for facilities that generate or manage hazardous waste, and requires a permit program for hazardous waste treatment, storage, and disposal facilities (TSDFs).[2] The basic goals of RCRA are the following: (1) to protect human health and the environment; (2) to conserve energy and natural resources; (3) to reduce waste generated, including hazardous and toxic wastes; and (4) to ensure effective and safe management practices to protect the environment. The EPA has the authority to enforce RCRA provisions to meet these objectives.

Among the major provisions of the act, hazardous wastes are defined as any solid waste or a combination of solid wastes, which due to its quantity, concentration, or physical, chemical, or infectious characteristics could (i) cause or significantly contribute to an increase in mortality or an increase in serious irreversible illness, or (ii) pose a substantial present or potential future hazard to human health or the environment when improperly treated, stored, transported, or disposed of, or otherwise mismanaged (Chapman, 1998: 174). Furthermore, wastes are considered hazardous whenever they exhibit or display any one or more of the attributes of ignitability, reactivity, corrosivity, and toxicity (as previously defined). RCRA also requires the tracking and record keeping of hazardous wastes from the point of generation, through transportation to storage, disposal, or treatment. Under RCRA provisions, hazardous waste generators are required to comply with regulations concerning record keeping and reporting; proper labeling of wastes; the use of appropriate containers; the release of information concerning the wastes' general chemical composition to transporters, handlers, processors, and disposers; and the use of a manifest system. All TSDFs are required to obtain permits from the EPA, comply with operating standards specified in the permit, meet financial requirements in case of accidents, and comply with RCRA provisions and guidelines. The states are also required to develop hazardous waste management plans to be approved by the EPA.[3]

Despite the breadth and depth of coverage of RCRA, loopholes continue to exist in toxic and hazardous wastes releases. For instance, there are a number of small-quantity waste generators not covered by the law. As noted by Jain et al. (1993: 27), regulation of waste oil, groundwater contamination due to leaking underground storage tanks (LUSTs), household generators, and toxic waste releases in agriculture remains outside the purview of this law. Also, as previously mentioned, the EPA has been criticized for its failure to enforce the RCRA provisions to curb exports of toxic e-waste.

HAZARDOUS AND SOLID WASTE AMENDMENTS (HSWA) OF 1984

Perhaps the most significant set of amendments to RCRA was the Hazardous and Solid Waste Amendments (HSWA) of 1984, which represents a complex law with several detailed technical requirements. Along with restrictions on land disposal and the inclusion of small-quantity hazardous waste generators (i.e., those producing 100–1,000 kg of waste per month) in the hazardous waste regulatory scheme, HSWA established the new regulatory program for underground storage tanks. This law also imposed on EPA a timetable for issuing or denying permits for treatment, storage, and disposal facilities; requiring permits to be for a fixed term not to exceed 10 years; and the required permit applications to be accompanied by information concerning the potential for public exposure to hazardous substances associated with the facility; it also gave the EPA the authority to issue experimental permits for facilities demonstrating new technologies. Other provisions of the HSWA prohibited the export of hazardous waste unless the government of the importing country formally agrees to accept it.[4]

THE COMPREHENSIVE ENVIRONMENTAL RESPONSE, COMPENSATION AND LIABILITY ACT

The CERCLA or the Superfund was passed in 1980 following the community contamination disaster at Love Canal. This legislation authorizes governmental responses to actual and potential releases of toxic and hazardous materials. Parties responsible for releases of hazardous substances may be held liable without regard to fault for certain damages caused by release, which basically include costs of clean-up, removal, and remediation incurred by the government. Originally, the CERCLA set up a hazardous substances response fund with an amount of $1.6 billion, which was increased to $8.5 billion with the

Superfund Amendments and Reauthorization Act (SARA) of 1986 (see Plater et al., 1992). An additional $500 million fund was also created for cleaning up leaking underground petroleum tanks. The petrochemical industry contributes the lion's share of the fund with the remaining coming from federal government revenues and other entities. The specific major provisions of the CERCLA are:

1. To set up a hazardous substance Superfund based on fees from industry and federal appropriations to finance response activities.
2. To establish liability to make up for the costs of response from liable or potentially responsible parties (PRPs), or to enforce the cleanup of sites by the responsible entity if identifiable.
3. To determine the number of inactive hazardous waste sites by conducting a national inventory. The EPA is required to work in conjunction with the Agency for Toxic Substances and Disease Registry (ATSDR) and the Public Health Service in the determination of the National Priority List (NPL) and implementation of health-related programs.
4. To authorize the EPA to respond to a release or a threat of a release of toxic pollutants from a site, which may adversely affect public health.

Prior to the passage of the CERCLA, the federal government had little power to actively respond to the growing problems of hazardous substance spills or contamination of sites with toxic and hazardous chemical substances (Kubasek and Silverman, 2000: 255). With the CERCLA, the federal government, through the EPA, was provided the necessary authority and funding mechanism to respond to these types of problems across the United States. Clearly, the CERCLA and its amendments (i.e., SARA) are more comprehensive, encompassing RCRA hazardous wastes and hazardous substances designated under the Clean Air Act, Clean Water Act, and the TSCA. Furthermore, the EPA is required to maintain an additional list of harmful substances not covered in the previous legislation. SARA required the use of the hazard ranking system (HRS) to determine the placement of sites on the NPL.

THE EMERGENCY PLANNING AND COMMUNITY RIGHT-TO-KNOW ACT (EPCRA)

The EPCRA passed in 1986 sets the requirements and a framework to direct the EPA, state and local governments, and the private sector to coordinate efforts to control and, when necessary, respond to releases of hazardous chemicals to the environment. It appears that the sudden, accidental release of methylisocyanate (MIC) to the environment at the Union Carbide plant in India in December 1984, which killed thousands of people and caused extensive chronic injuries, motivated many members of the Congress to support legislation to reduce the risk of chemical accidents within the United States. (Schierow, 2009). This act was designed to promote emergency planning to minimize the effects of toxic chemical accidents, and to provide information to the public concerning the nature and characteristics of chemical releases in communities. All state and local governments are required to have emergency response and preparedness plans. Each state is also required to have an emergency response commission, and each commission must appoint local emergency response planning committees and designate the emergency planning districts.

Section 304 of the act requires that facilities must immediately report a release of any "extremely hazardous substance" or any hazardous substance—a much broader category of chemicals defined under the CERCLA Section 102(a)—which exceeds the reportable quantity, to appropriate state, local, and federal officials.[5] The releases of a reportable quantity of a "hazardous substance" must also be reported to the National Response Center. EPCRA has led to a consistent reporting of toxic releases by industry and states across the country since 1988. The Toxics Release Inventory (TRI) database—a computerized EPA database of toxic chemical releases to the environment by manufacturing facilities—was established under the EPCRA.[6] Subtitle B of the act establishes various reporting requirements for facilities; those that manufacture, use, or process "toxic chemicals" are required to give an annual report to the EPA about the amounts of each chemical released into various environmental medium (i.e., air, land, and water) or the amounts transferred off sites. The data collected may be used to develop and implement emergency plans as well as furnish the public with general information about chemicals to which they may be exposed. The EPA makes TRI data available to the general public in special downloadable data files.

As described in the Congressional Research Service report by Schierow (2009: 2), the Occupational Health and Safety Act (OSHA) of 1970 requires most employers to give employees access to a material safety data sheet for any "hazardous chemical" being used. This right-to-know law for employees aims to ensure that people potentially exposed to such chemicals have free access to information about the potential adverse health effects of exposure and how to avoid them.

As Kubasek and Silverman (2000/2005: 307) note, new legislation and regulation after the passage of the EPCRA expand on prevention, response, and disclosure requirements. For instance, the Clean Air Act amendments of 1990 require many facilities to prepare risk management plans that effectively supplement the EPCRA emergency planning provisions. Furthermore, toxic release inventories have been expanded by the requirements of the Pollution Prevention Act of 1990 to cover recycling and source reduction activities.

In general, all of the US Federal legislation summarized in this section and in table 7.1 have produced significant outcomes in terms of reduction of toxic emissions in the country. For instance, from 1988 to 1999, total toxic releases on- and off-sites have decreased by 45.5 percent (a reduction of 1.46 billion pounds) (see EPA, 2000: E-7); also, from 2001 to 2007, total disposal or other toxic releases decreased by 27 percent.[7] Despite these legislations, however, the problems of hazardous and toxic waste management persist in the country and around the globe. The United States remains the largest generator of waste in the world, producing over 275 million tons annually. Consequently, the country has been slow or reluctant to regulate hazardous waste exports.

EUROPEAN COMMUNITY/UNION REGULATIONS

In response to the huge volume of wastes generated annually by the European Community (EC)/(European Union, EU) member states and disposed elsewhere, several directives underlying the regulation of hazardous waste have been introduced dating back to the mid-1970s. The 1975 Directive on General Principles of Waste Disposal, the 1978 Directive on Toxic and Dangerous Wastes, and the 1984 Directive on Trans-frontier Movement of Hazardous Wastes are briefly summarized in this section. Also, the recent directives on Waste Electrical and Electronic Equipment (WEEE) and the Restriction of Hazardous Substances (RoHS) will be revisited.

Among the provisions of the 1975 directive is the requirement for member states to make sure that wastes are disposed of without adverse effects on human health or the environment. Furthermore, EC member states are required to develop disposal plans, acceptable systems of waste transportation, storage and disposal facilities, and a method of tracking the movement of hazardous wastes.[8] The 1978 directive listed 27 generic types of waste, and it requires all member states to ensure proper transportation, treatment, and disposal of these wastes. However, this directive allowed each member state to define hazardous waste and to develop its own rule and regulatory system. This lack of harmonization of the definition of hazardous waste coupled with nonuniformity of the regulatory system, represents a major weakness of the directive.

The 1984 Directive on the Trans-frontier Movement of Hazardous Waste within the EC and waste exports to other states outside the EC, establishes and mandates a compulsory system of notification and tracking of hazardous waste movements. This directive was amended in 1986 to grant export of waste to other states outside the EU or EC only after the receiving or the importing nation has agreed to accept it and demonstrated a capacity to handle such waste. Thus, before any member state of the EU may export a hazardous waste, the holder must first notify the importing or receiving and transit nations of the shipment. Notification is provided through a uniform consignment note, similar to the manifest system used in the United States.[9]

More recently, two directives governing WEEE and the content of such devices were adopted by the EU after several years of debate. The directive on the WEEE and the directive on the RoHS became effective on February 13, 2003. Each member of the EU was required to implement these directives by means of laws, regulations, or administrative actions by August 13, 2004. As discussed in chapter 4, the overarching aim of the WEEE and RoHS directives is to substantially decrease the amount of e-waste entering incinerators and landfills, and to reduce or eliminate the hazardous substances these materials contain so as to protect human health and the environment.

INTERNATIONAL REGULATIONS OF HAZARDOUS WASTE AND TRANSBOUNDARY INDUSTRIAL ACCIDENTS

The systems of international regulations of transnational movements of hazardous wastes are recent developments. As indicated by Kubasek and Silverman (2005: 408), there is no international super-legislature to promulgate international laws, and an effective international mechanism with enforcement authority is lacking. The United Nations (UN), through its Security Council, is the only system attempting to meet the challenges of establishing and enforcing international regimes. Through the UN, international laws are derived from two key sources: treaties and other agreements freely signed and ratified by nation states and principles from long-standing practices. Norms derived from treaties are called conventional laws; and those based on long-standing practices are referred to as customary norms.

The Rotterdam Convention and the Basel Convention are important international conventional norms directed at regulating hazardous substances and transboundary movements of hazardous wastes, respectively.

THE ROTTERDAM CONVENTION

The growth in chemical production and international flows over the past five decades has raised international concerns about the potential risks posed by toxic chemicals, especially pesticides. Underdeveloped countries without adequate infrastructure to monitor the import and proper use of these chemicals are especially vulnerable. As a response to the mounting international concerns, the United Nations Environmental Program (UNEP) and the Food and Agriculture Organization (FAO) embarked upon developing and promoting voluntary information-exchange programs in the mid-1980s. By 1989, the two organizations introduced the Prior Informed Consent (PIC) procedure. Countries met in 1992 for the Rio Summit, where their officials adopted Chapter 19 of Agenda 21, which called for the adoption of a legally binding instrument on the PIC procedure by the year 2000 (UNEP, 2008). The FAO and the UNEP began negotiations of such an instrument, with the conclusion, adoption, and signing of the treaty in Rotterdam, the Netherlands, on September 10, 1998, which entered into force on February 24, 2004. Thus, this treaty called the Rotterdam Convention on Prior Informed Consent Procedure for certain Hazardous Chemicals and Pesticides in International Trade, establishes a legally binding instrument, by requiring parties to report their banned chemicals to the PIC Secretariat, which subsequently performs similar functions as the FAO/UNEP. The convention requires that any country importing pesticides and certain other hazardous or toxic chemicals must be informed of bans or severe restrictions on those chemicals in other countries. This essentially provides a receiving country the opportunity to refuse shipment of chemicals listed under the treaty on the grounds that they may pose significant threat to human health or the environment.

As stated in Article 1 of the Rotterdam Convention, the basic objective of this convention is to promote shared responsibility and cooperative efforts among parties in the international trade of certain hazardous chemicals in order to protect human health and the environment from potential harm and to contribute to their environmentally sound use, by facilitating information exchange about their characteristics. As of September 15, 2006, 109 countries and one regional economic integration organization had signed, ratified, accepted, approved, or acceded to the convention.[10] The United States remains the only outlier among advanced industrialized countries by refusing to ratify the Rotterdam Convention. Schafer (2006: 1) suggests that the inability of the United States to adopt or ratify the Rotterdam Convention and the Stockholm Convention constitutes a failure not just of US leadership in global environmental policy but of responsible participation in global efforts to protect human health as well.

THE BASEL CONVENTION

The Basel Convention on the Control of Trans-boundary Movements of Hazardous Wastes and Their Disposal represents the most comprehensive attempt at regulating hazardous wastes internationally. This convention came into force in 1989 under the aegis of the UNEP in response to growing concerns about toxic wastes from the more developed industrialized nations being dumped overtly and covertly in less developed countries (LDCs) of the global South and in nations with transitional economies. The discovery of illegal hazardous waste dumps in several parts of Africa (such as in the case of Koko discussed in chapter 6, the Khian Sea voyage around the globe, and radioactive wastes dumped in Benin) and in other parts of the developing world prompted international public outcry, which influenced the negotiations of a treaty to ban or restrict such practices. Rising costs of waste disposal; vanishing disposal sites due to Not in My Backyard (NIMBY) opposition; stringent regulations and enforcement within the United States, Western Europe, and other industrialized nations motivated seeking the paths of least or no resistance; and weak governance in underdeveloped countries led to hazardous waste dumping. This practice known as "toxic colonialism" or "toxic imperialism" has been condemned by the international community. As mentioned previously in chapter 6, the major driving force behind the acceptance of hazardous wastes in many impoverished nations has been the desperate need to earn income in the form of hard US dollars and EU currency. In some cases, greed among business establishments, the pervasive culture of bribery and corruption, ignorance of the nature of wastes being accepted, and low public awareness of waste dumping are also implicated.

Understandably, many African countries, supported by other developing nations and environmental nongovernmental organizations (NGOs), pushed for a total ban on transboundary shipment of hazardous waste as the only measure capable of forcing the more developed industrialized nations of the global North to dispose of their own hazardous waste and prevent them from using the impoverished underdeveloped nations of the global South as their dumping grounds. This was consistent with the Bamako Convention—a regional treaty negotiated subsequently by members of the Organization of African Unity (OAU). As expected, the concept of a total ban was vehemently opposed by the industrialized countries, especially the United States and the European Union, contending that the prohibition of trade in hazardous wastes having economic value, for example, in the recycling industry, between countries with adequate disposal options, would not be in the best interest of the environment (Kummer, 1998). Furthermore, these countries suggested that a total ban would violate the General Agreement on Trade and Tariffs (GATT) rules.

A growing demand by the global community for an international hazardous waste movement regulation culminated in the negotiations for the Basel Convention, which began in 1987. The final Conference of Plenipotentiaries (COP) on the Control of Transboundary Movements of Hazardous Wastes and Their Disposal was organized and convened in Basel, Switzerland, from March 20–22, 1989, to consider the final draft of the convention, mostly based on the Cairo Guidelines and Principles for the Sound Management of Hazardous Wastes, which offered recommendations concerning the export of hazardous wastes. These guidelines, contrary to a total ban, called for notification to receiving and transit countries of any waste export and consent by these countries prior to export. The waste exporter is required to make sure that the disposal site is adequate and capable of handling the hazardous waste and that the disposal complies with requirements at least as stringent as those applicable in the exporting country.

The Basel Convention was adopted unanimously by the conference on March 22, 1989 and 105 states and the EC signed the final act of the conference. The convention entered into force on May 5, 1992. It deals with hazardous wastes defined by means of a set of technical lists provided in its Annexes I and III; hazardous wastes are substances—toxic, poisonous, corrosive, explosive, flammable, infectious, and harmful to the environment—that are disposed of or are intended to be disposed of or are required to be disposed of by the provisions of national law.[11] Since its adoption, the number of parties to the convention has increased. As of June 2010, the total number of parties in the convention has reached 173 nations. Only the United

States, Canada, Japan, Australia, and New Zealand, among advanced the industrialized nations, and Afghanistan and Haiti, among the developing nations, have failed to ratify the convention.

Among the objectives of the convention are: (a) the restriction of transboundary movements of hazardous wastes except where it is perceived to be consistent with the principles of sound waste management; (b) the reduction of hazardous waste generation and treatment and disposal of hazardous wastes as close as possible to where they were generated; and (c) a regulatory system applicable to cases where transboundary movements of wastes are allowed (see Kummer, 1998). Thus, the basic aims of the convention are to minimize the generation of hazardous wastes and to control and decrease their transfrontier movements in order to protect human health and the environment.[12]

Prominent among the provisions of the convention is the requirement for PIC. Every country that is party to the convention may elect to ban the importation of hazardous wastes. With regard to other parties to the convention that have not banned waste imports, the country of export is required to assure prior notification of the governments of the importing or receiving state and any transit states in advance of any waste shipment. Article 6 of the Convention states:

> The state of export shall notify, or shall require the generator or exporter to notify, in writing, through the channel of the competent authority of the state of export, the competent authority of the states concerned of any proposed trans-boundary movement of hazardous wastes or other wastes. Such notification shall contain the declarations and information specified in Annex VA, written in a language acceptable to the state of import. The state of import shall respond to the notifier in writing, consenting to the movement with or without conditions, denying permission for the movement, or requesting additional information. The state of export shall not allow the generator or exporter to commence the shipment until it has received written confirmation that the notifier has received the written consent of the state of import; and the notifier has received from the state of import confirmation of the existence of a contract between the exporter and the disposer specifying environmentally sound management of the wastes in question.[13]

Thus, the convention prohibits the export of any hazardous or other waste until the importing states have given clear written authorization.

The consent of the proposed states notwithstanding, the Basel Convention further requires that states of export prohibit shipments of hazardous and other wastes if there is any reason to believe that the wastes will not be managed in an environmentally sound manner in the receiving country. The convention further articulates an obligation for states of export to ensure that transnational shipments of wastes are accepted for re-import if those shipments fail to conform to the terms of export (Wirth, 1998).

At the Conference of Plenipotentiaries (COP) held in March, 1994, the parties adopted the decision to immediately ban trade in hazardous waste destined for final disposal between developed countries and underdeveloped nations, and also to phase out trade in hazardous waste intended for reuse and recycling between these entities by December 2007.[14] Since this decision was not added to the text of the convention, there was a discord as to whether it was legally binding on the parties (Andrews, 2009). At the COP held in 1995, the parties adopted by consensus an amendment to the convention—"the Ban Amendment." This Ban Amendment had not been ratified by three quarters of the parties to the convention as required. Consequently, as of August 2009, only 65 out of 172 parties to the convention had ratified the Ban Amendment.[15]

A number of researchers have assessed the achievements and limitations of the Basel Convention (see Hackett, 1989; Krueger, 2001; Wirth, 1998; Andrews, 2009; Sende, 2010). While the accomplishments of the convention are quite remarkable in terms of establishing the first international norm to regulate the movements of hazardous waste across frontiers and galvanizing international support for this norm, several limitations have hampered the extent to which it could meet its original objectives. As Krueger (2001) notes, the convention has helped to put political pressure on hazardous waste-exporting countries in the global North; and it has put international spotlight on the problem of North-South toxic waste dumping and the need to stop it. But clear indicators of the effectiveness of the Basel Convention remain scanty due to the lack of consistent, comparable, and reliable data regarding the total volumes of hazardous waste traded across nations. The convention also lacks a rigorous monitoring and enforcement apparatus. Among the unintended consequences of the convention is the growing proliferation of the underground trade in hazardous waste from the global North to the global South. Some toxic wastes are often disguised as either material inputs, recycling materials, or comingled with other useful materials; and in some cases, hazardous wastes are labeled falsely.

Evidence of continued illicit toxic waste dumping in an African state despite the Basel Convention was presented in the case of 500 tons of toxic waste dumped in Abidjan, Ivory Coast (*Côte d'Ivoire*), in 2006. On August 19, a Panamanian registered ship, the *Probo Koala*, chartered by a Dutch company, unloaded the wastes at the port of Abidjan, and they were subsequently transported by tankers and dumped in 11 various unsecured locations within the city. The wastes consisted of liquid sludge containing large quantities of hydrocarbons, contaminated with at least hydrogen sulfide, mercaptans, and caustic soda. The wastes eventually leached into groundwater supplies, contaminating the city's drinking water, soil, and seafood. Local authorities reported more than 10 deaths and over 100,000 people sustaining injuries or other adverse health effects due to exposure to the toxic wastes.[16] Other recent cases of illegal toxic waste dumping in African states include Congo, Equatorial Guinea, and Somalia. In late 1998, more than 3,000 tons of mercury-contaminated industrial waste was found in an open dump near Sihanoukville, Cambodia. The waste, disguised and mislabeled as "cement cake" on import documents from Taiwan, was later implicated in local rioting that culminated in the deaths of at least two people. This waste was later returned to Taiwan.

THE CONVENTION ON THE TRANSBOUNDARY EFFECTS OF INDUSTRIAL ACCIDENTS

In response to a growing number of catastrophic industrial accidents with transboundary impacts, the United Nations Economic Commission for Europe (UNECE) organized a conference in Helsinki, Finland, in March 1992, focusing on developing an international framework for addressing the transboundary effects of industrial accidents. On March 17–18, 1992, 26 countries, including 14 member states of the EU, convened, deliberated, and signed the Convention on the Transboundary Effects of Industrial Accidents. This convention lays down a set of measures to safeguard human beings and protect the environment against the adverse effects of industrial accidents, and also to promote active international cooperation between the contracting parties before, during the impact, postimpact, and in the recovery phases of such accidents.

The convention applies to industrial accidents capable of causing transboundary effects, including accidents caused by natural disasters with the exception of the following:

- nuclear accidents or radiological emergencies;
- accidents at military bases or installations;

- dam failures with the exception of the effects of industrial accidents caused by such failures;
- land-based transport accidents with the exception of (a) emergency response to such accidents and (b) transportation on the site of the hazardous activity;
- accidental releases of genetically modified organisms (GMOs);
- accidents caused by activities in the marine environment, including seabed exploration or exploitation; and
- the spillage of oil and harmful substances at sea.[17]

It stipulates that all signatory states to the convention must take appropriate measures to prevent industrial accidents. Specifically, they must (1) induce action by operators to reduce the risk of industrial accidents; (2) establish policies or guidelines on the siting of new hazardous activities and on significant modifications to existing hazardous activities, with the aim of minimizing the risk to the population and the environment; and (3) be prepared for emergencies caused by industrial accidents, and be ready to introduce the necessary measures, including contingency plans to prevent, minimize, or mitigate transboundary impacts. Under this norm, the contracting states are required to ensure that adequate information is disseminated to the public in the areas susceptible to industrial accident arising out of a hazardous activity of transboundary magnitude.

Furthermore, the contracting states, in appropriate cases, must give the public an opportunity to participate in the decision-making process concerning prevention and preparedness initiatives. Natural or legal persons who are, or may be, affected by the transboundary effects of an industrial accident in the territory of a signatory state must be given the same access to the relevant administrative and judicial proceedings as a citizen of the state in question.[18] All parties to the convention must introduce a system of notification. In any event or imminent threat of an industrial accident with transboundary effects, the state of origin must notify the affected nations without delay and ensure that the response measures are taken as quickly as possible to contain and minimize the impacts of the accident.[19]

Article 1 of the Convention provides clear and concise definitions of the key terms. An industrial accident was defined as an event precipitated by an uncontrolled development in the course of any activity involving hazardous substances either in an installation, for instance, during manufacture, use, storage, handling, or disposal, or during transportation, insofar as it is covered by Paragraph 2(d) of Article 2. Hazardous activity refers to any activity in which one or more hazardous substances are present or may be present in quantities at or in excess of the threshold quantities listed in Annex I, which is capable of causing transboundary effects; and "effects" are conceptualized as any direct, acute, chronic, or delayed adverse consequences caused by an industrial accident on people, other organisms (i.e., flora and fauna), the biophysical environment (air, soil, water, and landscape), interaction between human population and environment, and material assets and cultural heritage, including historical monuments. The transboundary effects refer to serious impacts or consequences occurring within the jurisdiction of one party as a result of an industrial accident that occurred within the jurisdiction of another party.

Unlike the Basel Convention, parties to the convention on the transboundary effects of industrial accidents are exclusively from industrialized nations of the global North, among which there are disagreements, leading to a failure to reach a unanimous decision to fully ratify this convention. Conspicuously absent are the members of underdeveloped and developing societies of the global South. India and China, both of which have experienced one of the worst industrial disasters in history, are not parties to the convention. Consequently, the degree to which this convention can effectively meet its aims remains debatable.

CHAPTER SUMMARY AND CONCLUSIONS

Although significant progress has been made in legislative responses to environmental pollution in general, and toxic waste problems in particular since the publication of *Silent Spring* by Rachel Carson, there is still a long way to go to protect human health from heterogeneous toxic chemicals being produced and released, at such an alarming rate both in the United States and in other advanced industrial countries, and being moved throughout the planet. These toxic elements are being produced and released at a pace faster than that of the enactment of laws that are supposed to regulate them. Many new chemicals are persistent, deadly, and transboundary in nature requiring international laws and cooperation to regulate them. To this end, several international norms in the forms of global conventions and regional multilateral agreements have been forged to specifically deal with transboundary movements of hazardous waste, toxic chemicals, and industrial accidents.

The precautionary principle is now being advocated within the EU, the United States, and across the globe. This principle suggests that whenever there is a scientific uncertainty about the safety or potentially serious harm from chemicals or technologies, manufacturers or decision makers shall do everything possible to prevent harm to humans and the environment. In other words, "it is better to be safe than to be sorry," and manufacturers of toxic chemicals should be held accountable for any latent or manifest serious adverse health effects of these chemicals to humans and the environment. In the spirit of this principle, the national, regional, and global efforts to control, regulate, or ban the movement of hazardous substances to safeguard human health and to protect other species including the biophysical environment have been demonstrated in all the legislations and international accords presented in this chapter. Inasmuch as the problems seem to be growing and becoming more complex and challenging, new legislation, amendments to existing ones at the local level, and concerted international efforts to forge more effective international laws and develop more mechanisms to enforce existing ones would be required.

PART IV

CONCLUSION

In cases after cases of toxic waste releases, hazardous waste dumping, locally undesirable land uses (LULUs), and industrial disasters disadvantaged low-income minorities often bear a disproportionate brunt of the burden. It has been recognized that the distribution of environmental risks or hazards are not random. Pollution and other hazards tend to follow the path of least resistance as indicated by Robert D. Bullard. However, through grassroots mobilization of social capital and other resources, minority communities in the United States have orchestrated a powerful environmental justice movement (EJM) as an alternative to the mainstream environmental movement. Originally regarded as a fad, the EJM now has a lengthy milestone in its relatively short history. Environmental injustice is linked to most cases of community contamination. The victims often endure prolonged struggle for social and environmental justice. However, there are success stories of past environmental injustice cases being corrected throughout the country. In this concluding segment of the book, the theme of critical environmental justice struggles is addressed. The birth, incipiency, growth, and institutionalization of the EJM in the United States are covered. The key milestones in the evolution of EJM are discussed along with the growth of international EJM networks. Achievements till date and challenges that lie ahead for EJM are emphasized.

8

Conclusion: Critical Environmental Justice Struggles

Introduction

Environmental quality, locally undesirable land uses (LULUs), active and inactive hazardous industrial structures, landfills, and waste-to-energy installations are not evenly distributed in communities across the United States. Environmental inequality mirrors existing patterns of socioeconomic and racial inequality; race and class tend to be associated with living on the right track or on the wrong tract of the city's landscape.[1] Environmental hazards are physical conditions and circumstances in surroundings that may cause harm to the resident population, and the risks are the probability or likelihood of a hazard causing injury to an individual, a group, or a population (Fitzpatrick and LaGory, 2000: 10, 109; Douglas and Wildavski, 1982; Lupton, 1999). There is growing evidence that disadvantaged communities and populations of color are disparately exposed to a wide range of health-impairing and life-shortening hazards in their localities (Edelstein, 1988; Bullard, 1990; Fitzpatrick and LaGory; 2000; Bryant and Mohai, 1992).

Increased awareness of environmental-quality deprivation involving disparate exposure of African Americans, Latinos, Native Americans, and other people of color to health-impairing and life-threatening environmental hazards has ignited critical environmental justice struggles over the past 30 years. This closing chapter is devoted to issues of environmental justice both within the United States and around the globe. First, some conceptual definitions under the rubric of environmental justice are addressed. Different forms of environmental justice or injustice are discussed. Subsequently, the roots, evolution, institutionalization, and proliferation of the EJM within the country and across nations are discussed. The key events or milestones in the history of the movement, especially within the United States, are emphasized.

CONCEPTUAL DEFINITIONS

Environmental justice is the notion that no one should bear a disproportionate share of environmental hazards, especially if not directly involved in the industry or enterprise generating the hazards (see Bullard, 1990; Adeola, 1994; Taylor, 2000). The US Environmental Protection Agency (EPA) defines it as the fair and meaningful involvement of all people regardless of race, color, national origin, or income with respect to the development, implementation, and enforcement of environmental laws, regulations, and policies.[2] What is fair and what constitutes the meaning of involvement remain subjective. Nevertheless, this represents an ideal principle in sharp contrast to the real, everyday occurrences of locally unwanted land use (LULU) imposition on low-income minority communities as well as the unwarranted exposure of the poor and people of color to environmental hazards and associated health risks across the United States and beyond.

Different categories of environmental justice have been identified in the extant literature including distributive justice, procedural justice, and corrective or commutative justice (see Taylor, 2000; Kuehn, 2000; Widawsky, 2008). Distributive justice addresses the pattern of distribution of environmental "benefits" or "goods," and "environmental hazards," as well as social, economic, and political resources among communities with unequal levels of development and other outcomes. Procedural or representative justice on the other hand, is concerned with unequal bargaining powers and lack of adequate representation of members of communities of color in vital decision processes concerning environmental benefits and harms. It is widely recognized that the poor and racial/ethnic minorities are very often excluded or marginalized in decisions concerning LULUs and the placement of hazardous facilities (Bullard, 1990; Mohai and Bryant, 1992; Faber, 2008). The ultimate goal of procedural or representative justice is to ensure that a diversity of interests along racial, ethnic, and socioeconomic lines is adequately represented in the development and implementation of environmental agendas. Corrective or commutative justice calls for fair treatment of individuals, groups, or communities in environmental matters. It entails fairness in the manner in which punishments for lawbreaking are assigned and damages caused to individuals and communities are handled (Kuehn, 2000). Thus, it calls for "just compensation" for the victims of environmental injustice. Any individuals or groups that are made to bear the burdens of environmental hazards as a result of where they live, work, or recreate, should be duly compensated for by the responsible party if justice is to be served. Corrective justice aligns itself with the "polluter pays" principle and the "precautionary" principle.

An interesting question under the notion of restorative justice, however, is whether the victims of environmental injustice can be made whole again. While punitive sanctions can be imposed on identifiable perpetrators of environmental abuse, and financial compensation and reclamation measures can be used to correct past environmental injustices, it is rather difficult or impossible to make the victims whole again. Ostensibly, thousands of victims of the Bhopal gas disaster who died cannot be made whole again; thousands of victims of the Minamata mercury-poisoning episode who died cannot be made whole again; and many victims of environmental injustices along the Cancer Corridor of Louisiana with chronic health conditions induced by decades of environmental injustice cannot be made whole again by any compensatory award. In some cases, the judgments delivered are too little, and too late. Therefore, strict adherence to precautionary principles seems to be the most important means of preventing adverse health impacts.

Both the Executive Order 12898 signed by President Bill Clinton and the Principles of Environmental Justice (PEJ) (in Appendix 1) reflect the notions of restorative or corrective justice. The former directs federal government agencies to develop strategies to promote enforcement of health and environmental statutes in communities of color and low-income populations, and to collect, maintain, and evaluate data on race, national origin, and income of populations surrounding hazardous facilities or sites subject to federal enforcement action (see note 9). PEJ has a strong corrective justice scheme, asking for all toxic and hazardous waste producers to be held strictly accountable for cleanup and detoxification, and defining environmental justice as protecting the basic rights of victims of environmental justice to receive full compensation and reparations for damages, as well as proper health care.

Environmental injustice as an opposite term, manifests itself in various forms across sociodemographic characteristics—that is, race, ethnicity, class, rural-urban residency, and across nation states. It occurs within the society and between nations all over the world. From Love Canal, New York, to Bhopal, India, and from Koko, Nigeria, to Guiyu, China, the working class, the poor, and the minority groups are subject to exposure to environmental risks created by the operations of powerful affluent groups. As indicated by Beck (1999: 5), pollution generally follows the poor, who are the least able to mobilize resources to resist corporate pollution and other undesirable land uses. Environmental inequity involves unequal distribution of actual or potential sources of environmental hazards and their adverse health consequences among communities or segments of the population, in which one group bears the burdens (or negative externalities) more than the others, due largely to racial or socioeconomic differences. Environmental equity is the notion that environmental benefits and protection should be evenly distributed, while environmental hazards should be minimized or eliminated for all segments of the population.

Environmental racism, a term first used by the Rev. Benjamin Chavis, and its corollary, "environmental discrimination," feature prominently in environmental justice literature (see Bryant and Mohai, 1992; Bullard, 1990; Cole and Foster, 2001). Sociologically, the term racism involves negative, prejudicial attitudes and beliefs about a particular subordinated race/ethnicity often used as the basis for ill-treatment of the members of the group. The promotion of ideological racism (the belief that one race is biologically or innately intellectually and culturally superior to others) has major consequences. When members of the dominant race claim to be superior to all other races by virtue of the power and privilege they possess, racism becomes woven into the fabric of a culture. It pervades every sphere of social life, and therefore, it is not by accident that environmental racism is a fact of life. Racist ideology has been used to justify the domination, colonization, exploitation, and imposition of environmental burdens on those races considered inferior.[3] Environmental racism can be defined as any environmental decisions, actions, and policies that disadvantage or harm any racial/ethnic minority community (either intentionally or unintentionally based on race). Environmental racism/discrimination stems from three interrelated factors including: (a) the existence of negative prejudicial beliefs and actions; (b) having the power to formulate and implement policies and actions that reflect one's own racial prejudices and stereotypes; and (c) privileged, unfair opportunities over other racial/ethnic groups and the capacity to advance one's group interests over another.[4]

The concept of environmental racism has been employed in describing a wide range of situations adversely affecting the people of color, including the following:

- having one or more LULUs and other noxious facilities in one's community or backyard;
- the increased chance or probability of being exposed to environmental hazards where one lives, works, learns, or plays;

- disparate negative effects of environmental policies; for example, unequal rates of environmental law enforcement and fines, and differential rates or speed of cleanup of contaminated waste sites in communities of color relative to predominantly white communities;
- deliberately targeting and placing of LULUs and other noxious facilities in low-income, minority communities;
- environmental blackmail—occurring when people are forced to choose between dangerous employment or environmental standards;
- segregation of racial/ethnic minority workers in hazardous and dirty occupations;
- lack of access or restricted access to or substandard maintenance of environmental amenities;[5]
- increased chances of environmentally induced morbidity and mortality among racial/ethnic minority populations; and
- First World industrialized nations deliberately targeting other nations, especially the less developed countries (LDCs) of the global South for hazardous waste dumping.

According to Bullard (1993: 98), racism plays a key role in most environmental planning and decision making across the United States. In fact, it is reinforced by vital institutions such as the government, legal, economic, political, and military structures. Environmental racism is juxtaposed with the public policies and industry practices of allocating environmental benefits or amenities to whites, while shifting the bulk of negative externalities to communities of people of color (see Bullard, 2000: 98; Faber, 2008).

Critical environmental justice struggle becomes inevitable among the marginalized and disenfranchised groups when they are confronted with health-impairing and life-threatening environmental hazards considered unjustly imposed by one or more outsiders. The demand of fundamental rights to live in a clean, healthy, safe, and productive environment by communities whose rights are infringed upon, or whose environment has been contaminated by big corporations, governments, or other powerful entities, is the essence of critical environmental justice struggle. Many cases presented in this book such as the Bhopal gas disaster, the contamination episodes at Love Canal, New York, Koko, Nigeria, and Abidjan, Côte d'Ivoire, and the *maquiladoras* along the US-Mexico border communities have demonstrated that environmental justice is often a matter of life or death. Human life and well-being are sacrificed in the course of environmental exploitation for the sake of wealth accumulation, and paradoxically, human life and freedom are often sacrificed in the course of critical environmental justice struggles, especially in societies of the global South.

When compared to other mainstream environmental movements, EJM is relatively new, emerging in the early 1980s. The amalgamation or fusion of the civil rights movement, antitoxic, social justice, occupational- and environmental health movements, and the indigenous land struggles by Native Americans, African Americans, and Chicanos, and human rights and EJMs have been addressed in the literature asserting that the latter borrowed some of the strategies and tactics employed by the civil rights activists (Sarokin and Schulkin, 1994; Faber, 2008: 224). As Cole and Foster (2001) suggest, identifying a particular date or event that gave birth to the EJM is problematic because of the fact that this movement evolved from scores or hundreds of other local struggles that predate the 1950s, and that it learned from the movements of the 1960s and the 1970s. For instance, within the United States, Native Americans and other people of color have engaged in critical environmental justice struggles over access to land and natural resources for centuries. Ceasar Chavez led the protests of the United Farm Workers against exploitation and exposure to toxic pesticides in agricultural fields in California's San Joaquin valley in the 1960s.

Also, for many centuries, many parts of the global South have struggled against colonialism, imperialism, apartheid, and coercive usurpation, exploitation, and degradation of their environments by nations or multinational corporations (MNCs) of the global North. The history of environmental injustice and flagrant environmental racism is replete with instances such as the following: spraying toxic chemicals such as Agent Orange on native peoples and their ecosystem; the subjugation of Native Americans and their relocation to reservations with diminished rights over natural resources; continuous use of their lands as a "national sacrifice zone" for military weapons testing, leaving behind several unexploded ordinances, and for disposal of nuclear wastes; the historical enslavement of blacks and continuous placement of hazardous or toxic waste facilities in their backyards; and the intentional and deliberate targeting of Third World nations for toxic waste exports (Hooks and Smith, 2004; Pellow, 2007; Greenpeace, 1994; Clapp, 2001; Faber, 2008). As Robert D. Bullard points out, right from the outset, institutionalized racism pervaded the economic, political, social, and ecological spheres that support the subjugation and exploitation of subordinate groups and the environment.

In his book, *Capitalizing on Environmental Justice: The Polluter-Industrial Complex in the Age of Globalization*, Daniel Faber describes the global North-global South patterns of unequal exchange and environmental injustice in the colonization of people and nature under the guise of globalization. From the political economy standpoint, the economic prowess of global North capitalism is inextricably dependent on the appropriation of surplus value and ecological space from the global South. The major role of corporate-propelled globalization and the neoliberal agenda is to facilitate and accelerate the exploitation of labor and withdrawal of cheaper raw materials and consumer goods from the global South to the global North while using the former as a repository of hazardous wastes generated. An increased level of awareness and a growing sensitivity to environmental racism in the United States ignited collective grassroots movements with the aim of mitigating environmental injustice.

THE ORIGINS, EVOLUTION, AND DIFFUSION OF THE ENVIRONMENTAL JUSTICE MOVEMENT

The birth of the EJM has been linked to the protests against the contamination of the middle-class neighborhood of Love Canal, New York, in the late 1970s, discussed previously in chapter 6. Originally founded as an antitoxic movement that grew out of the Love Canal Homeowners Association, the Citizens Clearing House for Hazardous Wastes organized by Lois Marie Gibbs, who was a young mother and housewife at the time, first focused attention on class disparity in community toxic exposures and associated health problems. It has since shifted its focus and mission to educating and helping local people throughout the country to fight for environmental justice, and has been renamed the Center for Health, Environment, and Justice.[6] The concept of environmental racism was not the master frame employed in the Love Canal struggle, albeit it demanded commutative or corrective justice for all the afflicted residents. Through "popular epidemiology" (the process through which lay residents uncovered the sources of disease patterns afflicting their community), accompanied by a shared understanding of the problems' sources and solutions, the risk, antitoxic, and community contamination frames emerged (Levin, 1982; Brown, 1992; Brown and Mikkelsen, 1990; Blum, 2008).

Several scholars identified the protest and civil disobedience over the dumping of toxic polychlorinated biphenyl (PCB) waste in Afton, a predominantly African American community in Warren County, North Carolina, in 1982, as the origin of the EJM in the United States (Cole and Foster, 2001; Lester et al., 2001; Bullard, 1990; McGurty, 2009; Johnson, 2009). As pointed out by Rhodes (2003: 58), minority communities have faced a series of environmental threats long before the 1970s and 1980s, and have protested against LULUs, occupational hazards, workers' health and safety issues, pesticide contamination, and location of facilities considered hazardous to community health and well-being, but to no avail. By the late 1970s and early 1980s, however, several structural alterations in the American society brought environmental issues and the problems of environmental injustice to the limelight.

Around the period when the Love Canal saga was unfolding, a number of low-income communities of color were being targeted for toxic waste dumping or for hosting LULUs. During the summer of 1978, a company dealing with waste breached the Toxic Substances Control Act (TSCA) by engaging in "mid-night dumping" of PCB-contaminated liquid waste removed from the Ward Transformer Company in Raleigh, North Carolina, onto the roadside, covering and contaminating more than a 240-mile stretch of soil. This was done to avoid the high cost of PCB disposal under the existing regulation. The state of North Carolina subsequently assumed the responsibility of cleaning up the contaminated roadside, and chose the small community of Afton as the site for a new landfill to contain the PCB-contaminated soil. The specific site was previously a farmland bought at a cheap price from a distressed farmer who was facing foreclosure and bankruptcy.[7] Public announcement of the new disposal site immediately ignited intense resistance from county residents who happened to be predominantly African Americans. They expressed concerns about a possible contamination of underground water supply and a potential threat to community development due to the stigma associated with a toxic waste facility.

After all the legal recourses were exhausted, residents proceeded to launch collective civil disobedience to disrupt activities at the landfill site. According to McGurty (2009: 4), the citizens were supported by regional and national civil rights leaders as they organized protest events daily, borrowing the tactics and strategies previously used by civil rights activists. The protesters lay down on the road to block the passage of dump-trucks carrying PCB-contaminated soil to be disposed of at the Afton dump site. Most poignantly, racism was invoked as the underlying motivation for locating a PCB dump site in the predominantly African American rural community of Afton in

Warren County. This protest led to the arrest and jailing of more than 500 people, including a prominent member of the US Congress, former Congressman Walter Fauntroy. Even though this protest failed to deter the state government from implementing its action, it succeeded in calling national attention to the problems of environmental injustice/racism and ignited the EJM. In other words, EJM is fundamentally a targeted response to patterns of environmental inequities and the corresponding injustices. One specific outcome of the events was a study conducted in 1983 by the US General Accounting Office (GAO) in all EPA Region IV hazardous waste sites. In the study, race and income were found to be correlated with the location of hazardous waste facilities. The study revealed that 75 percent or 3 out of 4 of the off-site, commercial hazardous waste facilities in the studied region were placed in predominantly African American and poor communities (GAO, 1983).

Another landmark study "Toxic Waste and Race in the United States," was conducted in 1987 by the United Church of Christ's Commission for Racial Justice (CRJ) based on a national systematic sample. The study popularized the term "environmental racism," a term coined by Benjamin Chavis of the United Church of Christ as previously mentioned. Similar to the GAO report, race was found as the most significant variable factor in predicting where commercial and public waste treatment and disposal facilities were sited. The findings of these studies were consistent with the patterns uncovered by sociologist, Robert D. Bullard, in his studies of Houston, Texas, Cancer Alley, Louisiana, and the entire Deep South.[8] Subsequent empirical research by sociologists and other social scientists found similar evidence in different regions, states, and localities throughout the United States (see Adeola, 1994, 2000; Bryant and Mohai, 1992; Mohai and Bryant, 1992; Novotny, 2000; Bullard, 1990; Lester et al., 2001; Mohai and Saha, 2007).

The Michigan coalition—consisting of a group of academicians, researchers, and activists, led by Bunyan Bryant and Paul Mohai—convened at a meeting held on the campus of the University of Michigan in January, 1990. At the conference, empirical evidence and policy solutions and directions concerning disproportionate distribution of environmental burdens by race and socioeconomic standings were discussed. Among the outcomes of the conference was the correspondence and subsequent meetings with William Reilley, EPA administrator, and the recognition of environmental justice as a legitimate issue to be addressed by the agency. In 1992, the EPA administrator, William Reilley, established the Office of Environmental

Equity, during the administration of republican president, George H. W. Bush. It was subsequently renamed Office of Environmental Justice during the Clinton administration. In a sense, EJM was cata-pulted to the national stage, marking the onset of the institution-alization of the movement. Another important development in the early 1990s was the publication of *Dumping in Dixie*, by Robert D. Bullard, which systematically documented the significance of race in toxic facilities distribution and exposures to health-impairing hazards in the Deep South.

Perhaps one of the most significant milestones in the evolution of the EJM was the First People of Color Environmental Leadership Summit held in Washington, DC, in October 1991. This summit attracted more than 1,000 attendees from across the United States and from around the globe. More than 600 delegates from 50 states of the United States were present. Besides forging regional, national, and global EJM networks, one significant product of this summit was the crafting and introduction of the principles of environmen-tal justice (PEJ). A set of 17 universal principles reflecting the civil rights norms and the universal human rights declarations, to serve as a guide for developing and evaluating policies and programs for environmental and social justice, was released to the global commu-nity (see Appendix I). This was unprecedented in the history of social movements. As mentioned by Sze and London (2008: 1334), the PEJ provided an alternative framework for environmentalism by exceed-ing the racial, ethnic, gender, and class biases in mainstream envi-ronmental groups, and by combating the abuses of both corporate polluters and government regulatory agencies.

In a special issue of the *National Law Journal*, Lavell and Coyle (1992) revealed that there is unequal protection and enforcement of existing environmental laws by the EPA. Enforcement of regulations is relaxed or often ignored, and when implemented, fines for viola-tions are extremely low for predominantly white communities relative to those for communities of color. The growing scholarly interest and accumulation of volumes of credible empirical research findings, revealing different patterns of environmental inequity and credible and substantiated evidence of environmental injustice, were among the catalysts that further energized the coalescence of a full-fledged EJM in the early 1990s. As is the case with most social movements, opposing views and critical research also evolved leading to healthy debates among academicians, social scientists, and environmental jus-tice activists (see Been, 1994, 1995; Anderton et al, 1994; Yandle and Burton, 1996; Mohai and Saha, 2007; Zimmerman, 1993). The opposing views often challenged or disputed the environmental racism claim with emphasis placed on economic inequalities—that is, environmental burdens are the inescapable by-products of market forces in which minority people seek affordable places located in pollution-prone landscapes; or the location of hazardous facilities are guided by rational forces of supply and demand. These opposing views notwithstanding, millions of people of different racial, ethnic, cultural, and national backgrounds now recognize that environmental discrimination is unacceptable, immoral, unethical, and unjustifiable.

At this time, the EJM has evolved and consists of civil rights groups, antitoxic groups, occupational and environmental health groups, religious groups, farm workers, tenant associations, professional not-for-profit groups, academics, labor unions, university research centers, and several others (Agyeman, 2005). Supporters and sympathizers have been growing in leaps and bounds within the American public. In fact, the movement has stretched across the country—from the Southern to the Northern states and from West coast to East coast and to all other regions—being driven by the strong activism of multiethnic and multiracial activists employing the master frame of environmental justice and passionately committed to the core objective of achieving environmental justice for all. Furthermore, environmental justice has been recognized as a legitimate field of inquiry and a course of study is being offered at several colleges and universities across the United States (Figueroa, 2002).

Other major historical events that shaped the EJM are summarized in table 8.1. The asterisked events are considered the forerunner milestones setting the stage for the birth and growth of EJM in the subsequent years. Efforts have been made by the liberals to codify environmental justice legislation, especially Rep. John Lewis (D-GA) and Sen. Al Gore (D-TN) who jointly introduced the Environmental Justice Act in 1992; however, strong opposition from the conservatives led to the defeat of this legislation. Subsequent attempts to reintroduce the bill also failed. It was not until 1994 that President Bill Clinton signed the Executive Order 12898, directing all federal agencies to identify, prioritize, and address disproportionately high negative health and environmental effects of their policies, programs, and activities on low-income people and minority people of color.[9]

Undoubtedly, the Executive Order 12898 has empowered numerous grassroots groups and minority communities throughout the United States from the 1990s to the present. It has served as an impetus to the cleaning and revitalization of several brownfields in low-income communities across the United States. Many environmental

Table 8.1 Major Milestones in the Evolution of Environmental Justice Movement

Date	Key Event(s)
Early 1960s*	Farm workers organized by Cesar Chavez protested for workplace rights, including safety and protection from toxic pesticide exposures in California agricultural fields.
1962*	Rachel Carson's *Silent Spring* was published, which chronicles the harmful effects of pesticides on organisms and the environment.
1964*	The Civil Rights Act was passed. The Title VI of the law prohibits the use of federal funds to discriminate based on race, color, and national origin.
1970*	The US Environmental Protection Agency was set up to enforce existing laws that protect human health and the natural environment.
1971*	President's Council on Environmental Quality confirms that racial discrimination negatively affects the quality of the environment of the urban poor.
1972*	The United States bans the use of DDT, one of the most toxic pesticides.
1978	The government evacuated hundreds of families from Love Canal, New York, due to contamination by toxic chemicals buried several decades earlier under the community.
1979	African American community in Houston, Texas, opposing a landfill, filed the first Title VI lawsuit challenging the siting of the waste facility.
1982	Predominantly African American community in Warren County, North Carolina, protests against dumping PCB-contaminated soil in the neighborhood; this represents the first nationally recognized environmental protest by African Americans.
1983	The US General Accounting Office (GAO) released the report of a study entitled "Siting of Hazardous Landfills and Their Correlation with Racial and Economic Status of Surrounding Communities," in which it found three-fourths of the total number of hazardous waste-disposal sites located in poor African American communities.
1984	Accidental release of toxic gas (MIC) from a pesticide-manufacturing plant owned by the US-based Union Carbide killed more than 6,000 people.
1987	The United Church of Christ's Commission for Racial Justice released a report entitled "Toxic Wastes and Race in the United States," which shows that race is the most significant factor in predicting where toxic waste facilities are sited in the country.
1988	Britain's Black Environment Network was established (http://www.ben-network.org.uk/).
1989	The Basel Convention on the Control of Trans-boundary Movements of Hazardous Wastes and Their Disposal was adopted and signed by 105 states and the European Community on March 22.
1990	Robert D. Bullard published *Dumping in Dixie: Race, Class, and Environmental Quality*, which underscores the significance of race in siting undesirable toxic-releasing facilities. The Michigan Conference was held in January, 1990, under the leadership of Bunyan Bryant and Paul Mohai at the University of Michigan's School of Natural Resources and Environment, assembling teams of academics and activists to discuss the empirical evidence and policy solutions concerning disproportionate environmental burdens. The Michigan Coalition wrote letters and met with William Reilly, EPA administrator, in September.
1991	The First National People of Color Environmental Leadership Summit convenes in Washington, DC, in October, with more than 1,000 attendees and establishes the guiding principles of environmental justice.

Continued

Table 8.1 Continued

Date	Key Event(s)
1992	Rep. John Lewis (D-GA) and Sen. Al Gore (D-TN) jointly introduced the Environmental Justice Act. It failed to pass.
	National Law Journal publishes a special issue on Unequal Protection that chronicles the double standards and disparate treatment of people of color and whites.
1994	President Bill Clinton signs the Executive Order 12898, which directs federal agencies to identify and address disproportionately high adverse health and environmental effects of their policies or programs on low-income people and people of color.
1997	Citizens against Nuclear Trash (CANT) and local residents in Houma, Louisiana, win a major victory over Louisiana Energy Services (LES) on Earth Day. CANT blocks the LES from building a uranium-enriched plant in the middle of Forest Grove and Center Springs, Louisiana.
2002	Second People of Color Environmental Leadership Summit or EJ Summit II was convened on October 24–27 in Washington, DC, with over 1,400 participants in attendance.
2003	Cleanup or detoxification of the Warren County, North Carolina, PCB landfill was completed at a cost of 17.1 million dollars.
2005	Hurricane Katrina disaster exposes existing environmental injustice and human deprivation in New Orleans and the Deep South.
	The General Accounting Office (now renamed the US Government Accountability Office) releases a report documenting that EPA generally devotes little attention to environmental justice issues while drafting three significant clean air rules, on gasoline, diesel, and ozone, between 2000 and 2004.
	A December 2005 study from the Associated Press (AP) finds that more blacks live with pollution. The AP reported that black Americans are 79% more likely than whites to live in neighborhoods where industrial pollution is suspected to pose the greatest health threat.
2006	The Concerned Citizens of Agriculture Street Landfill in New Orleans wins a major legal victory after 13 years of litigation demanding relocation and buyout from their contaminated community. In 2005, Hurricane Katrina created a forced relocation of the residents; however, in January 2006, five months after Katrina, 17th District Court Judge, Nadine Ramsey, ruled in favor of the plaintiffs—describing them as overwhelmingly poor minority citizens who were promised the American dream of first-time home ownership, which turned out to be a nightmare.
2010	After 25 years of litigation, a court in Bhopal, India, found the Union Carbide India Limited and eight former officials guilty of causing death by negligence, culpable homicide, and gross negligence under the Indian Penal Code. The officials were each sentenced to two years imprisonment and a fine of approximately $2,097, and the company was fined approximately $10,483.

Source: Adapted from various sources including: Bullard, R. D., P. Mohai, R. Saha, and B. Wright. *Toxic Wastes and Race at Twenty, 1987-2007: A Report Prepared for the United Church of Christ Justice & Witness Ministries*. Cleveland, OH: United Church of Christ, 2007, pp.16–37, available at http://www.ucc.org/assets/pdfs/toxic 20.pdf (accessed August 19, 2010); Summit II National Office, Environmental Justice Timeline-Milestones, available at http://www.ejrc.cau.edu/summit2/%20EJTimeline.pdf (accessed August 18, 2010); and Natural Resources Defense Council, The Environmental Justice Movement, available at http://www.nrdc.org/ej/history/hej.asp (accessed August 18, 2010); GAO. "Environmental Justice: EPA Should Devote More Attention to Environmental Justice When Developing Clean Air Rules," available at http://www.gao.gov/new.items/d05289.pdf (accessed August 21, 2010).

justice legal battles have been fought and notable victories have been reported including the following: prevention of Shintech Corporation from building the polyvinyl chloride (PVC) plant in Convent, Louisiana, a predominantly African American Community; the blockage of a new nuclear-enrichment plant in Forest Grove and Center Springs, Louisiana; cleanup and detoxification of the PCB landfill in Warren County, North Carolina; the buyout and relocation of residents of Diamond community, demarcated by a fence-line from the polluting Shell Oil Company's chemical plant and refinery in Norco, Louisiana;[10] and, the rulings for the plaintiffs of the Agriculture Street Landfill in New Orleans, Louisiana, after 13 years of legal battle, that provided just compensation. There are numerous other similar cases and outcomes throughout the United States.

As presented in table 8.1, another remarkable milestone in the evolution of EJM was the second National People of Color Environmental Leadership Summit (Summit II) held in Washington, DC, in October 2002. This event attracted a huge number of participants from all over the globe. As stated by Bullard (2005: 22), more than 14,000 individuals representing grassroots and community-based organizations, faith-based groups, organized labor, civil rights groups, youth groups, and academicians/researchers were present. Participants came from different continents including Africa, Europe, Asia, North America, and South and Central America. About 20 different nations were represented by delegates in attendance at the summit, giving credence to the transnational scope the EJM has achieved.

Since the 1990s, EJM has also grown across the globe with extensive involvement of international networks and nongovernmental organizations (NGOs) such as the Indigenous Environmental Network, Greenpeace International, Pesticide Action Network, Basel Action Network (BAN), and several others assisting in facilitating connections between grassroots groups, and between organizations, and empowering them to build resistance against environmental and social injustices. To a significant extent, the process of globalization, Internet access and connectivity, and accelerated telecommunication systems have enhanced global networking among various grassroots activist organizations, including EJMs. As observed by Zavestoski (2009), besides fractionalization and conflicts among grassroots groups, local efforts to connect with the international community to establish global networks to promote environmental and social justice in Bhopal were unsuccessful until in the early 2000s. The International Campaign for Justice in Bhopal (ICJB) was formed in 2002 (nearly 18 years after the tragedy) out of the vestiges of defunct or unsuccessful networks and coalitions of the late 1980s through the 1990s. The ICJB currently consists of a coalition of individuals, organizations, nonprofit groups, and other entities that have mobilized resources and joined forces to actively and relentlessly campaign for justice for the survivors and victims of the Union Carbide India Limited (UCIL) disaster in Bhopal, India. Members of the ICJB continue to press the Dow Chemical Company (who acquired UCIL) and the US and Indian governments to guarantee sufficient current and future health-care rehabilitation, a safer environment, and an equitable compensation for the survivors of the disaster. The ICJB embraces and promotes five basic principles including: the precautionary principle, the polluter pays principle, the community right-to-know principle, international liability, and environmental justice.[11]

One specific example in the UK often viewed as a case of an environmental justice struggle was the plight of the community of Greengairs in Lanarkshire, Scotland. Similar to Warren County, North Carolina, Greengairs was targeted as a dumping ground for PCB-contaminated soil collected from across the UK. A grassroots environmental justice campaign supported by the Friends of the Earth, Scotland, was launched to stop the toxic assault on the community (Pedersen, 2010). Other notable milestones in the development of EJM around the globe include the establishment of the Black Environment Network (BEN) in Britain in 1988; the Basel Convention on the Control of Trans-boundary Movements of Hazardous Wastes and Their Disposal, adopted by 105 nation states and the European community in 1989, and its subsequent ban amendment, which regulates international dumping of toxic wastes—an aspect of global environmental injustice; the emergence of environmental justice groups in other parts of the global North and the diffusion of EJM to other areas in the global South; the compensation of Bhopal victims; and the subsequent rulings in June, 2010, after 25 years of legal struggles, finding the UCIL and its officials guilty of gross negligence and culpable homicide with fines and jail terms. Whether the rulings represent justice or injustice remains open to debate, which is beyond the scope of this chapter.

There are numerous examples illustrating that EJM has become a phenomenon to be taken seriously in developing countries of the global South. For instance, in Colombia, starting in 1992, indigenous people organized and protested against a drilling project by Occidental Petroleum until the project was stopped in 2002; similarly in Brazil, a landless Workers' Movement became energized and active in large-scale land acquisitions; and in the Niger Delta of Nigeria, the Ogoni people waged a protracted struggle against environmental injustice and continuous exploitation of their homeland by the Royal Dutch Shell, a multinational company extracting millions of barrels of oil without due compensation to the people. The Movement for the Survival of the Ogoni People (MOSOP)—organized under the leadership of the late Ken Saro-Wiwa—aimed to achieve political autonomy, environmental and social justice, and prevention of ecological degradation of the Ogoniland. The transnational outreach, resource mobilization, and spotlight accorded MOSOP became a threat to the regime of a dictator. In 1995, Ken Saro-Wiwa and eight other activists were summarily put to death by the Abacha administration (see Adeola, 2000, 2001, 2009; Okonta and Douglas, 2001; Staggenborg, 2011; Comfort, 2002). There are several others like Saro-Wiwa and Chico Mendes who have paid the ultimate price in the struggle for environmental justice.

CHAPTER SUMMARY

By critical environmental justice struggles, I mean the enduring confrontation to reclaim the rights to clean air, water, land, and a productive environment capable of supporting life and the well-being of all people. While it has always being the practice for the most powerful group or class to impose hazards such as LULUs, industrial plants and their effluents, and major accidents on low-class minorities, it has been demonstrated that through collective efforts and the use of appropriate strategies and tactics of planned change, minority groups can be empowered to confront giant corporations and governmental agencies at the roots of environmental injustice. For the communities facing environmental contamination due to racial discrimination in toxic waste dumping, the fight for environmental justice is a matter of life and death (Bullard, 2005; Auyero and Swistun, 2008). Several case studies presented in the previous chapters of this book vividly attest to this statement, especially the cases of Love Canal, New York; Minamata, Japan; Guiyu, China; Bhopal, India; Koko, Nigeria; Woburn, Massachusetts; and Agriculture Street Landfill, New Orleans. As mentioned, environmental racism and environmental injustice are immoral, unethical, and downright inhumane as they are tantamount to violation of fundamental human rights.

Despite the setbacks during the eight years of George W. Bush administration, grassroots organization and activism persist at the local, regional, and national levels within the United States. One among the factors associated with the EJM setbacks during the Bush administration was documented in the report by the GAO (formerly Government Accounting Office) in 2005 indicating that the EPA devoted little attention to environmental justice considerations while drafting the three important clean air rules on gasoline, diesel, and ozone, between the fiscal years 2000 and 2004. In terms of safeguarding cross-national environmental justice, the GAO also found the EPA to be deficient in controlling the exports and illegal shipments of toxic electronic waste (e-waste) as mentioned in chapter 4 (GAO, 2008; 2005).

EJM now appears ubiquitous—spreading throughout the United States, Canada, Europe, Africa, Asia, and Central and Latin America. Extensive international networks have been forged through the participation of delegates at major international conferences and summits. Environmental justice has been recognized as a civil rights and human rights issue enforceable under the national and international norms. This chapter has chronicled the brief history of the EJM in the United States with emphasis on key milestones in the evolution of the movement. The number of grassroots organizations has grown exponentially since 1982, and based upon the growing interest in environmental justice issues locally, nationally, and internationally, the future of this movement is quite promising. Much work lies ahead and several obstacles are yet to be surmounted. For instance, as noted by Faber (2008: 243), we must acknowledge the fact that unlike many mainstream environmental organizations, environmental justice non-profit organizations led by people of color are grossly undersupported by the philanthropic community in the United States. Courting the philanthropic community and other sources of resources are among the challenges facing the EJM in the near future.

Appendix I

Principles of Environmental Justice (PEJ)

Delegates to the first National People of Color Environmental Leadership Summit held between October 24 and 27, 1991, in Washington DC, drafted and adopted 17 principles of environmental justice. Since then, these principles have served as a defining document for the growing grassroots movement for environmental justice.

Preamble

WE, THE PEOPLE OF COLOR, gathered together at this multinational People of Color Environmental Leadership Summit, to begin to build a national and international movement of all peoples of color to fight the destruction and taking of our lands and communities, do hereby re-establish our spiritual interdependence to the sacredness of our Mother Earth; to respect and celebrate each of our cultures, languages and beliefs about the natural world and our roles in healing ourselves; to ensure environmental justice; to promote economic alternatives which would contribute to the development of environmentally safe livelihoods; and, to secure our political, economic and cultural liberation that has been denied for over 500 years of colonization and oppression, resulting in the poisoning of our communities and land and the genocide of our peoples, do affirm and adopt these Principles of Environmental Justice:

1. Environmental Justice affirms the sacredness of Mother Earth, ecological unity, and the interdependence of all species, and the right to be free from ecological destruction.
2. Environmental Justice demands that public policy be based on mutual respect and justice for all peoples, free from any form of discrimination or bias.
3. Environmental Justice mandates the right to ethical, balanced, and responsible uses of land and renewable resources in the interest of a sustainable planet for humans and other living things.

4. Environmental Justice calls for universal protection from nuclear testing, extraction, production, and disposal of toxic/hazardous wastes and poisons, and nuclear testing that threaten the fundamental right to clean air, land, water, and food.

5. Environmental Justice affirms the fundamental right to political, economic, cultural, and environmental self-determination of all peoples.

6. Environmental Justice demands the cessation of the production of all toxins, hazardous wastes, and radioactive materials, and that all past and current producers be held strictly accountable to the people for detoxification and the containment at the point of production.

7. Environmental Justice demands the right to participate as equal partners at every level of decision-making, including needs assessment, planning, implementation, enforcement, and evaluation.

8. Environmental Justice affirms the right of all workers to a safe and healthy work environment without being forced to choose between an unsafe livelihood and unemployment. It also affirms the right of those who work at home to be free from environmental hazards.

9. Environmental Justice protects the right of victims of environmental injustice to receive full compensation and reparations for damages as well as quality health care.

10. Environmental Justice considers governmental acts of environmental injustice a violation of international law, the Universal Declaration on Human Rights, and the United Nations Convention on Genocide.

11. Environmental Justice must recognize a special legal and natural relationship of Native Peoples to the US government through treaties, agreements, compacts, and covenants affirming sovereignty and self-determination.

12. Environmental Justice affirms the need for urban and rural ecological policies to clean up and rebuild our cities and rural areas in balance with nature, honoring the cultural integrity of all our communities, and provide fair access for all to the full range of resources.

13. Environmental Justice calls for the strict enforcement of principles of informed consent, and a halt to the testing of experimental reproductive and medical procedures and vaccinations on people of color.

14. Environmental Justice opposes the destructive operations of multinational corporations.

15. Environmental Justice opposes military occupation, repression, and exploitation of lands, peoples, and cultures, and other life forms.

16. Environmental Justice calls for the education of present and future generations which emphasizes social and environmental issues, based on our experience and an appreciation of our diverse cultural perspectives.

17. Environmental Justice requires that we, as individuals, make personal and consumer choices to consume as little of Mother Earth's resources and to produce as little waste as possible; and make the conscious decision to challenge and reprioritize our lifestyles to ensure the health of the natural world for present and future generations.

NOTES

1 SOCIOLOGY OF HAZARDOUS WASTES, DISASTERS, AND RISK

1. In his presidential address to the American Sociological Association, James F. Short called on fellow sociologists to conduct basic and applied "policy relevant" research on risks and disasters; see James F. Short, "The Social Fabric at Risk: Toward the Social Transformation of Risk Analysis," *American Sociological Review* 49 (1984): 711–725.
2. Many of these cases have been covered extensively in the literature. See Robert E. Hernan, *This Borrowed Earth: Lessons from the 15 Worst Environmental Disasters around the World* (New York: Palgrave/ Macmillan, 2010).
3. Hernan, *Borrowed Earth*, pp. 61–100; Phil Brown, *Toxic Exposures: Contested Illnesses and the Environmental Movement* (New York: Columbia University Press, 2007); Elizabeth D. Blum, *Love Canal Revisited: Race, Class, and Gender in Environmental Activism* (Lawrence: University Press of Kansas, 2008); Phil Brown and E. J. Mikkelson, *No Safe Place: Toxic Waste, Leukemia and Community Action* (Berkeley: University of California Press, 1990).
4. Robert D. Bullard, *Dumping in Dixie: Race, Class and Environmental Quality* (Boulder, CO: Westview Press, 1990); also see B. L. Allen, *Uneasy Alchemy: Citizens and Experts in Louisiana Chemical Corridor Disputes* (Cambridge: MIT Press, 2003).
5. See P. Brown, "Popular Epidemiology and Toxic Waste Contamination: Lay and Professional Ways of Knowing," *Journal of Health and Social Behavior* 3 (September, 1992): 267–281; Hernan, *Borrowed Earth*, pp. 61–100; M. R. Edelstein, *Contaminated Communities: The Social and Psychological Impacts of Residential Toxic Exposure* (Boulder, CO: Westview Press, 1988): 43–83.
6. US Department of Health and Human Services, Agency for Toxic Substances and Disease Registry (ATSDR), available at: http://

www.atsdr.cdc.dov/docs/APHA-ATSDR-book.pdf (accessed May 16, 2010).

2 Hazardous and Toxic Wastes as a Social Problem

1. C. W. Schmidt, "Unfair Trade: E-Waste in Africa," *Environmental Health Perspectives* 144, no. 4 (2006): A232–235; Jim Pucket et al., *The Digital Dump: Exporting Re-use and Abuse to Africa* (Seattle, WA: BAN, 2005).
2. Ulrich Beck, *World Risk Society* (Malden, MA: Polity Press, 1999).

3 Taxonomy of Hazardous Wastes

1. See http://www.atsdr.cdc.gov/HEC/caselead.html.(accessed December 19, 2009).
2. See http://www.niehs.nih.gov/health/topics/agents/mercury/index.cfm (accessed November 19, 2009).
3. EPA/FDA, "What You Need to Know About Mercury in Fish and Shellfish" (2004), http://www.fda.gov/food/foodsafety/product-pecificinformation/seafood/foodbornepathogenscontaminant:/Methymercury/UCM115662.htm (accessed December 19, 2009).
4. A. Comte and A. Flury-Herard, "Radioactive Waste: What Health Effects, or Risks?" *CLEFS CEA* 53 (Winter, 2005–2006): 9–11, http://www.cea.fr/var/cea/storage/static/gb/library/clefs53/pdf-gb/009-11pflury-53gb.pdf (accessed January 22, 2010).
5. Comte and Flüry-Herard, "Radioactive Waste."
6. See "Nuclear Information and Resource Service: Radioactive Waste Project, Yucca Mountain, Nevada: Proposed High-Level Radioactive Waste Dump Targeted at Native American Lands," http://www.nirs.org/radwaste/yucca/yuccahome.htm (accessed January, 22, 2010).
7. See US DOE, "Transuranic Waste Processing" (2009), http://www.becteljacobs.com/pdf/factsheets/TRU_Waste_fact_sheet.pdf (accessed January 23, 2010).
8. US EPA "Radiation: Risks and Realities" (2007), http://www.epa.gov/rpdweb00/docs/402-K-07-006.pdf (accessed January 22, 2010).
9. See The Chernobyl Forum, "Chernobyl's Legacy: Health, Environmental and Socioeconomic Impacts and Recommendations to the Governments of Belarus, the Russian Federation, and Ukraine" (2003–2005), http://www.iaea.org/Publications/Booklets/Chernobyl/Chernobyl.pdf (accessed January 23, 2010).

4 ELECTRONIC WASTE: THE DARK SIDE OF THE HIGH-TECH REVOLUTION

1. See "Electronics Take Back Coalition" (July 2008), available online at: http://computertakeback.com/tools/facts_and_figures.pdf (accessed December 27, 2009).
2. "Electronics Take Back Coalition, US" (July 2008).
3. Joe Walsh, "Global E-Waste Market Is Forecast to Reach 53 Million Tons by 2012 says New Report," *PR-Log-Global Press Release Distribution,* (http://www.prlog.org/10453267-global-ewaste-markets-is-forecast.html (accessed January 2, 2010).
4. Walsh, "Global E-Waste Market 2008–2012."
5. EPA, "Electronic Waste Management in the United States, Approach 1 Table 3.1 EPA530-R-08-009," http://www.epa.gov/osw/conserve/materials/ecycling/docs/app-1.pdf (accessed December 27, 2009).
6. SHOROC, "Fact Sheet 1. Electronic Waste in Australia—A growing problem," http://www.shoroc.nsw.gov.au/pdf/fact%sheet%201%20ElectronicElectronic%20Waste%20in%20Australia.pdf (accessed December 30, 2009).
7. See E. Williams and R. Kuehr, "Today's Markets for Used PCs and Ways to Enhance Them," in R. Kuehr and E. Williams (eds.) *Computers and the Environment: Understanding and Managing Their Impacts* (Tokyo, Japan: United Nations University Press, 2003), pp. 197–210.
8. Walsh, "Global E-Waste Market."
9. Heavy metals are defined as chemical elements with a specific gravity that is at least five times the specific gravity of water, which is 1 at 4°C (39°F). See http://www.lef.org/protocols/prtcl-156.shtml#comm (accessed February 17, 2010).
10. A Superfund site is any area in the United States that has been contaminated by hazardous waste and has been determined by the EPA as a candidate for cleanup due to the risk it poses to human health and the environment. Superfund was established by Congress through the provision of the Comprehensive Environmental Response, Compensation and Liability Act (CERCLA) to pay for cleanup or remediation of abandoned toxic waste sites across the country. The fund is financed by fees paid by toxic waste generators and by cost recovery from cleanup projects.
11. M. Eugster et al., "Sustainable Electronics and Electrical Equipment for China and the World: A Commodity Chain Sustainability Analysis for Key Chinese EEE Product Chains," Draft Working

Paper (IISD, November, 2007). Also, see W. Yang, "Regulating Electrical and Electronic Wastes in China," *RECIEL*, 17, no. 3 (2008): 337–346.

12. K. Brigden, I. Labunska, D. Santillo, and M. Allsopp, "Recycling Electronic Wastes in China and India: Workplace and Environmental Contamination," Technical Note 09/2005 (Section 1), Greenpeace International, Greenpeace Research Laboratories, Department of Biological Sciences, University of Exeter: Exeter, UK, 2005.

13. Eric Williams, R. Kahhat, B. Allenby, E. Kavazanjian, J. Kim, and M. Xu, "Environmental, Social, and Economic Implications of Global Reuse and Recycling of Personal Computers," *Environment, Science and Technology*, 42 (2008): 6446.

14. R. Kahhat and E. Williams, "Product or Waste? Importation and End-of-Life Processing of Computers in Peru," *Environment, Science and Technology*, 43 (2009): 6010.

15. Circuit boards removed from any product are also regulated under RCRA. Circuit boards intended to be disposed of, recycled, or reclaimed would come under the definition of hazardous waste. EPA regulations, however, provide a conditional exclusion from the hazardous waste definition for circuit boards that are shredded for recycling after removal of certain hazardous components, and an exemption from the definition of whole circuit boards to be recycled, which are considered scrap metal. These circuit boards are not subject to any regulatory requirements when exported.

16. "Basel Convention on the Control of Trans-boundary Movements of Hazardous Wastes and Their Disposal," available at http://www.basel.int/text/con-e-rev.pdf (accessed January 4, 2010).

17. "Basel Convention on the Control of Trans-boundary Movements of Hazardous Wastes and their Disposal: Meeting the Global Waste Challenge," available at http://www.basel.int/convention/about.html (accessed June 9, 2010).

18. "Basel Convention on the Control of Trans-boundary Movements of Hazardous Wastes."

19. European Commission. Recast of the WEEE and RoHS Directives Proposed, available at http://ec.europa.eu/environment/waste/weee/index_en.htm (accessed June 10, 2010). Also see INFORM, "European Union (EU) Electrical and Electronic Products Directives: Directive on Waste Electrical and Electronic Equipment (WEEE), Directive on the Restriction of the Use of Certain Hazardous Substances in Electrical and Electronic Equipment (RoHS)," http://www.informinc.org/weeover.pdf (accessed June 10, 2010).

5 ENVIRONMENTAL HEALTH RISKS OF PERSISTENT ORGANIC COMPOUNDS

1. Othmar Zeidler first synthesized DDT in Strasbourg, Germany, in 1874. In 1938, Edward Dodds announced the synthesis of DES, and a Swiss chemist, Paul Muller, also announced the insecticidal properties of DDT, hailed as a miracle, which earned him the Nobel Prize for Physiology and Medicine in 1948. It was first used to control typhus from lice in 1942 and subsequently, the US Department of Agriculture discovered that it was an effective exterminator of numerous pests including mosquitoes, flies, fleas, and other insects.
2. See the UNEP website at http://www.chem.unep.ch/sc/
3. A comprehensive or exhaustive list of adverse health effects of POPs and allied toxic chemical compounds is beyond the scope of the present study.
4. A list of all other food items included in the FDA study is available online at http://vm.cfsan.fad.gov/~dms/pes99rep.html
5. For technical notes, methods, and list of food items sampled in the states in the United States and foreign countries, see tables 1, 2, and 6 of FDA's *Total Diet Study* at the site in note 2.

6 COMMUNITIES CONTAMINATED BY TOXIC WASTES AND INDUSTRIAL DISASTERS: SELECTED CASES

1. See the World Disaster Report, 2007. The total number of reported technological disasters in the world increased from 201 to 371 in 2005. The trend in natural disasters is even much higher (see table 5 of the report).
2. Also, see Epstein et al., 1982, *Hazardous Waste in America*, and Blum, 2008, *Love Canal Revisited: Race, Class and Gender in Environmental Activism*.
3. Levine, 1982.
4. Eckard C. Beck, "The Love Canal Tragedy," *EPA Journal* (January, 1997), http://www.epa.gov/history/topics/lovecanal/01.htm (accessed May 17, 2010).
5. http://www2.shore.net/~dkennedy/woburn_mit.html (accessed November 20, 2009).
6. See David Hammer, "Court Upholds Dump Housing Payout," *The Times Picayune* (2008), http://www.nola.com/news/index. ssf/2008/06/court_upholds_dump_housing_pay.html (accessed May 16, 2010).

7. See the Louisiana Supreme Court ruling in Agriculture Street case, http://blog.nola.com/news_impact/2008/06/AGSTREET070108. pdf (accessed May 16, 2010).

8. See TED case studies on Bhopal at http://www.american.edu/ted /phopal.htm (accessed May 16, 2010).

9. See "Nigeria: N 39 million Relief for Koko Toxic Waste Victims 21 Years After," *Vanguard* (April 4, 2008), http://allafrica.com/stories/ 200804041094.html (accessed May 13, 2010); also, "Nigeria: Koko Toxic Waste Victims Get Compensation 20 yrs. After," *This Day* (April 4, 2008), http://allafrica.com/stories/200804040085.html (accessed May 13, 2010).

10. *Maquiladora* is a term derived from the Spanish word *maquilar*, which connotes the service a miller renders when he grinds wheat into flour. In similar fashion, a maquiladora provides assembly services without assuming ownership of the goods being assembled. Also see GOA, 2003.

11. See Environmental Health Coalition (EHC), http://www.envi-ronmentalhealth.org/Maquilapolis/maquiladoras.html (accessed December 19, 2010).

12. See Connie Garcia and A. Simpson, "Globalization at the Crossroads: Ten Years of NAFTA in the San Diego/Tijuana Border Region," *EHC* (2004), http://www.environmentalhealth.org/PDFs/PDFs_Archive /globalizationFNLREL_10_18_04.pdf (accessed December 19, 2010).

13. See EHC, note 11.

14. Environmental News Service, "Lead Waste Capped at Abandoned Smelter on US-Mexico Border" (2009), http://www.ens-newswire. com/ens/jan2009/2009-01-28-091.asp (accessed December 22, 2010).

7   The Regulatory Frameworks

1. A summary of US environmental regulations is provided at http://tis. eh.doe.gov/oepa/law_sum/TSCA.HTM (accessed May 17, 2010).

2. Summaries of environmental laws administered by the EPA are pro-vided by the Congressional Research Service (CRS) report RL 30022 http://www.nconline.org/nle/cresreports/briefingbooks/laws/h./ cfm (accessed July 18, 2010).

3. See the Congressional Research Service Report RL30022.

4. CRS report RL 30022, p. 8.

5. As defined under CERCLA section 102(a), a "hazardous substance" includes any element, compound, mixture, solution, and substance

that upon release into the environment may present significant or serious danger to the public health or welfare and the environment.

6. By definition, "toxic chemicals" are substances in solid, liquid, or gaseous forms that may cause serious harm or sickness among the population exposed to them, even in relatively small doses or amounts, by eating, drinking, breathing, or through dermal absorption. As previously explained in chapter 3, the term hazardous substance is broader, encompassing toxic chemicals and substances that are corrosive, explosive, flammable, or harmful.

7. US EPA, "Toxic Release Inventory, Reporting Year 2007, Public Data Release: Summary of Key Findings," http://www.epa.gov/TRI/trida-ta/tri07/pdr/key_findings_v12a.pdf (accessed October 27, 2009).

8. See Counsel Directive on General Principles of Waste Disposal, 18, O.J. EUR, Comm. (No L194), 39 (1975); Directive on Toxic and Dangerous Waste, 21 O.J., EUR. Comm. (No L 84), 43 (1978).

9. David P. Hackett, "An Assessment of the Basel Convention on the Control of Transboundary Movements of Hazardous Waste and their Disposal," *American University Journal of International Law and Policy* 5 (1989): 291–323.

10. See UNEP/FAO, "Status of Ratification of the Rotterdam Convention as of September 15, 2006, UNNEP/FAO/RC/COP.3/INF/1," http://www.pic.int/cops/cop3/yinf1)/English/k0652731%20cop-3-INF1.pdf (accessed August 2, 2010).

11. See the Basel Convention: http://www.basel.int/text/con-e.pdf (accessed August 3, 2010); also, see J. Krueger, "The Basel Convention and the International Trade in Hazardous Wastes," in *Yearbook of International Cooperation on Environment and Development*, ed. Olav SchramStokke and Oystein B. Tommessen (London: Earthscan 2001/2002), pp. 43–51.

12. The Basel Convention, note 11 above.

13. See Basel Convention Ratification at http://basel.int/ratif/convention.htm (accessed August 2, 2010), and note 11 above.

14. Decision II/12 in Report of the Second Meeting of the Conference of Parties to the Basel Convention on the Control of Transboundary Movements of Hazardous Waste and Their Disposal, UN Doc. UNEP/CHW, (1994), pp. 2–30.

15. Decision III/I, in Report of the Third Meeting of the Conference of Parties to the Basel Convention on the Control of Transboundary Movements of Hazardous Wastes and Their Disposal, UN Doc. UNEP/HW3/35 (1995).

16. There are varying reports on the number of casualties. These figures are obtained from a UNEP press release, "Donor Governments Should Support On-Going Cote d'Ivoire Emergency," December 14, 2006, http://www.unep.org/Documents.Multilingual/Default.asp? (accessed August 2, 2010).
17. Europa, Summaries of EU Legislation, Transboundary Effects of Industrial Accidents, http://www.europa.au/legislation_summaries/environment/civil_protection.html (accessed August 4, 2010); and Convention on the Transboundary Effects of Industrial Accidents, Treaty No 5, Presented to Parliament by the Secretary of State for Foreign and Commonwealth Affairs, February 2003, http://.fco.gov.uk/resources/en/pdf/pdf2/fco_pdf_cm5741pdf_cm5741_industrialaccid (accessed August 7, 2010).
18. Convention on the Transboundary Effects of Industrial Accidents, Treaty No. 5.
19. Convention on the Transboundary Effects of Industrial Accidents, see note 18.

8 Conclusion: Critical Environmental Justice Struggles

1. Factors associated with people ending up on the right or wrong track of a city's landscape are discussed by K. Fitzpatrick and M. Lagory, *Unhealthy Places: The Ecology of Risk in the Urban Landscape,* (New York: Routledge, 2000).
2. The US EPA, "Environmental Justice," available online at http://www.epa.gov/environmentaljustice/ (accessed August 15, 2010).
3. For an extensive discussion of "racism" and "racist ideology," see J. E. Farley, *Majority-Minority Relations*, 6th ed. (New York: Prentice Hall/Pearson, 2010); and R. T. Schaefer, *Racial and Ethnic Groups*, 12th ed. (Upper Saddle River, NJ: Prentice Hall/Pearson, 2010).
4. See D. Taylor, "Environmental Racism," Pollution Issues. http://www.pollutionissues.com/Ec-Fi/Environmental-racism.html (accessed August 14, 2010).
5. Taylor, Environmental Racism.
6. The Center for Health, Environment, and Justice, http://www.chej.org/ (accessed August 18, 2010).
7. See Eileen McGurty, *Transforming Environmentalism: Warrant County, PCBs, and the Origins of Environmental Justice* (New Brunswick, NJ: Rutgers University Press, 2009); "Warrant County, NC, and the Emergence of the Environmental Justice Movement:

Unlikely Coalitions and Shared Meanings in Local Collective Action," *Society and Natural Resources* 13, no. 4 (2000): 373–387.

8. Robert D. Bullard, *Dumping in Dixie: Race, Class, and Environmental Quality* (Boulder, CO: Westview Press, 1990, 2000).
9. William J. Clinton, Executive Order 12898 of February 11, 1994: Federal Actions to Address Environmental Justice in Minority Populations and Low-Income Populations. *Federal Register* 59, no. 32.
10. For a detained case study of Diamond, see S. Lerner, *Diamond: A Struggle for Environmental Justice in Louisiana's Chemical Corridor* (Cambridge: MIT Press).
11. The International Campaign for Justice in Bhopal (ICBJ) provides an overview of the group online at http://bhopal.net/oldsite/icjb.html (accessed August 12, 2010).

BIBLIOGRAPHY

Abelsohn, A., B. L. Gibson, M. D. Sanborn, and E. Weir. "Identifying and Managing Adverse Environmental Health Effects: Persistent Organic Pollutants." *Canadian Medical Association Journal* 166, no. 12 (2002): 1549–1554.

Adeola, Francis O. "Environmental Hazards, Health, and Racial Inequity in Hazardous Waste Distribution." *Environment and Behavior* 26 (1994): 99–126.

———. "Environmental Contamination, Public Hygiene, and Human Health Concerns in the Third World: The Case of Nigerian Environmentalism." *Environment and Behavior* 28, no. 5 (1996): 614–646.

———. "Cross-National Environmentalism Differentials: Empirical Evidence from Core and Noncore Nations." *Society and Natural Resources* 11, no. 4 (1998): 339–364.

———. "Endangered Community, Enduring People: Toxic Contamination, Health, and Adaptive Responses in a Local Context." *Environment and Behavior* 32, no. 2 (2000a): 209–249.

———. "Cross-National Environmental Injustice and Human Rights Issues: A Review of Evidence in the Developing World." *American Behavioral Scientist* 43, no. 4 (2000b): 686–705.

———. "Environmental Injustice and Human Rights Abuse: The States, MNCs, and Repression of Minority Groups in the World System." *Human Ecology Review* 8, no. 1 (2001): 39–59.

———. "Toxic Waste." pp. 146–177 in *Waste in Ecological Economics,* edited by K. Bisson and J. Proops. Cheltenham, UK: Edward Elgar Publishing, 2002.

———. "Boon or Bane? The Environmental and Health Impacts of Persistent Organic Pollutants." *Human Ecology Review* 11, no. 1 (2004): 27–35.

———. "From Colonialism to Internal Colonialism and Crude Socio-environmental Injustice: Anatomy of Violent Conflicts in the Niger Delta of Nigeria." pp. 135–163 in *Environmental Justice in the New Millennium: Global Perspectives on Race, Ethnicity, and Human Rights,* edited by Filomina C. Steady. New York: Palgrave MacMillan, 2009.

Agency for Toxic Substances and Disease Registry (ATSDR). "Case Studies in Environmental Medicine: Lead Toxicity." US Department of Health and Human Services. Atlanta, GA: ATSDR, 1990. http://www.astdr.cdc.gov/HEC/case lead.html (accessed October 10, 2009).

———. "Public Statement for Mercury." Atlanta, GA: ATSDR. 1990. http://www.atsdr.cdc.gov/ToxProfiles/phs8916.html. (accessed October 10, 2009).

———. "Public Health Statement for Lead." Atlanta, GA: ATSDR. 1997. http://www.atsdr.cdc.gov/toxprofiles/phs13/ (accessed December 10, 2009).

———. "CERCLA Priority List of Hazardous Substances." 2007. http://www.atsdr.cdc.gov/cercla/07list.html (accessed December 8, 2009).

———. "Toxicological Profile for Mercury (update)." Atlanta, GA: US Department of Health and Human Services, Public Health Service, Agency for Toxic Substances and Disease Registry. March, 1999. http://www.astdr.cdc.gov/ToxProfiles/tp46-c1-b.pdf (accessed October 10, 2009).

———. "Fact Sheet: Cadmium," CAS #7440-43-9, Atlanta, GA: ATSDR, 2008. http://www.atsdr.cdc.gov/tfacts5.pdf (accessed November 10, 2009).

Agyeman, Julian. "Sustainable Communities and the Challenge of Environmental Justice." New York: New York University Press, 2005.

Albrecht, S. L. and R. G. Amey. "Myth-Making, Moral Communities, and Policy Failure in Solving the Radioactive Waste Problem." *Society and Natural Resources* 12, no. 8 (1999): 741–761.

Alcock, R. E., M. S. McLachlan, A. E. Johnson, and K. C. Jones. "Evidence for the Presence of PCDD/Fs in the Environment Prior to 1900 and Further Studies on Their Temporal Trends." *Environment Science and Technology* 32, no. 11 (1998): 1580–1587.

———. "Response to Comment on, 'Evidence for the Presence of PCDD/Fs in the Environment Prior to 1900 and Further Studies on Their Temporal Trends.'" *Environment Science and Technology* 33, no. 1 (1999): 206–207.

Allchin, D. "The Poisoning of Minamata." *Ships Teacher's Network*. 1999. http://www2.utep.edu/~allchin/ships/ethics/minamata.htm (accessed October 8, 2009).

Allen, Barbara L. *Uneasy Alchemy: Citizens and Experts in Louisiana's Chemical Corridor Disputes.* Cambridge: MIT Press, 2003.

Amzal, B., B. Julin, M.Vahter, A. Wolk, G. Johanson, and A. Akesson. "Population Toxicokinetic Modeling of Cadmium for Health Risk

Assessment." *Environmental Health Perspectives* 117, no. 8 (2009): 1293–1301.

Anderton, Douglas L. Methodological Issues in the Spatiotemporal Analysis of Environmental Equity." *Social Science Quarterly* 77, no. 3 (1996): 508–515.

Anderton, D. L., Andy B. Anderson, J. M. Oates, and M. Fraser. "Environmental Equity: The Demographics of Dumping." *Demography* 31 (1994): 229–248.

Andrews, A. "Beyond the Ban—Can the Basel Convention Adequately Safeguard the Interests of the World's Poor in the International Trade of Hazardous Waste?" *Law Environment and Development Journal* 5, no. 2 (2009): 167–184.

Andrews, Anthony. "Radioactive Waste Streams: Waste Classification for Disposal." Congressional Research Service Report, December 13, 2006.

Arcury, T. A. and Sara A. Quandt. "Chronic Agricultural Chemical Exposure among Migrant and Seasonal Farmworkers." *Society and Natural Resources*, 11 no. 8 (1998): 829–843.

Asante-Duah, D. K. *Hazardous Waste Risk Assessment.* Boca Raton, FL: Lewis Publishers, 1993.

Auyero, Javier and Debora Swistun. "The Social Production of Toxic Uncertainty." *American Sociological Review* 73, no. 3 (2008): 357–379.

Babu, B. R., A. K. Parande, and C. A. Basha. "Electrical and Electronic Waste: A Global Environmental Problem." *Waste Management Research* 25, no. 4 (2007): 307–318.

Baccarelli, A., A. C. Pesatori, D. Consoni, P. Mocarelli, D. G. Patterson Jr., N. E. Caporaso, P. A. Bertazzi, and M. T. Landi. "The Health Status and Plasma Dioxin Levels in Chloracne Cases 20 Years after the Seveso, Italy, Accident." *British Journal of Dermatology* 152, no. 3 (2005): 459–465.

Bair, Scott E. and M. A. Metheny. "Remediation of the Wells G & H Superfund Site, Woburn, Massachusetts." *Ground Water* 40, no. 6 (2002): 657–668.

Baldassare, M. and C. Katz. "The Personal Threat of Environmental Problems as Predictor of Environmental Practices." *Environment and Behavior* 24 (1992): 602–616.

Ballschmiter, K., R. Hackenberg, W. M. Jarman, and R. Looser. "Man-made Chemicals Found in Remote Areas of the World: The Experimental Definition for POPs." *Environmental Science and Pollution Research* 9, no. 4 (2002): 274–288.

200   Bibliography

Basel Action Network and Silicon Valley Toxics Coalition. *High-Tech Trashing of Asia.* Seattle, WA: BAN, 2002.

Baskin, Laurence, Katherine Himes, and Theo Colborn. "Hypospadias and Endocrine Disruption: Is There a Connection? *Environmental Health Perspectives* 109, no. 11 (2001): 1175–1183.

Baum, Andrew. "Toxins, Technology and Natural Disaster." pp. 5–54 in *Cataclysms, Crises, and Catastrophes,* edited by G. Van de Bos. Washington, DC: American Psychological Association, 1987.

Baum, A. and I. Fleming. "Implications of Psychological Research on Stress and Technological Accidents." *American Psychologist* 48, no. 6 (1993): 665–672.

Baum, Andrew, I. Fleming, and J. E. Singer. "Coping With Victimization by Technological Disaster." *Journal of Social Issues* 39, no. 2 (1983): 117–138.

Baum, Andrew, I. Fleming, A. Isreal, and M. K. O'Keeffe. "Symptoms of Chronic Stress Following a Natural Disaster and Discovery of a Human Made Hazard." *Environment and Behavior* 24, no. 3 (1992): 347–365.

Beck, Ulrich. *Risk Society: Towards a New Modernity.* Newbury Park, CA: Sage Publications, 1992.

———. *World Risk Society.* Cambridge, MA: Polity, 1999.

———. *Ecological Enlightenment: Essays on the Politics of the Risk Society.* Atlantic Highlands, NJ: Humanities Press, 1995.

———. *World at Risk.* Cambridge, UK: Polity Press, 2007.

———. *Ecological Politics in an Age of Risk.* Cambridge, MA: Polity Press, 1995.

———. "World Risk Society as Cosmopolitan Society? Ecological Questions in a Framework of Manufactured Uncertainties." *Theory, Culture and Society* 13, no. 4 (1996): 1–32.

Been, Vicky. "Locally Undesirable Land Uses in Minority Neighborhoods: Disproportionate Siting or Market Dynamics?" *Yale Law Journal* 103, no. 6 (1994): 1383–1422.

———. "Analyzing Evidence of Environmental Justice." *Journal of Land Use and Environmental Law* 11 (1995): 1–36.

Berman, E. *Toxic Metals and Their Analysis.* Philadelphia, PA: Hayden and Son, Inc., 1980.

Bernard, A. 2008. "Cadmium and Its Adverse Effects on Human Health." *Indian Journal of Medical Research* 128, no. 4 (2008): 557–564.

Bertazzi, P. A., I. Bermucci, G. Brambilla, D. Consonni, and A. C. Pesatori. "The Seveso Studies on Early and Long-Term Effects of Dioxin Exposure: A Review." *Environmental Health Perspectives* 106, no. 52 (April 1998): 625–635.

Blackman Jr., and C. William. *Basic Hazardous Waste Management*. New York, NY: Lewis Publishers, 1996.

Blocker, T. J. and D. L. Eckberg. "Gender and Environmentalism: Results from the 1993 General Social Survey." *Social Science Quarterly* 78, no. 4 (1997): 841–858.

Blum, Elizabeth D. *Love Canal Revisited: Race, Class, and Gender in Environmental Activism*. Lawrence: University of Kansas Press, 2008.

Bodeen, C. "In E-Waste Heartland, a Toxic China." *New York Times*, November 18, 2007. http://www.nytimes.com/2007/11/18/World /Asia/18iht-Waste.1.837 (accessed April 4, 2010).

Braud, Karl-Werner. "Environmental Consciousness and Behavior: The Greening of Lifestyles." pp. 204–217 in *International Handbook of Environmental Sociology*, edited by M. Redclift and G. Woodgate. Northampton, MA: Edward Elgar, 1997.

Brigden, K., I. Labunska, D. Santillo, and P. Johnston. *Chemical Contamination at E-Waste recycling and Disposal Sites in Accra and Korforidua, Ghana*. Greenpeace Research Laboratories, Department of Biological Sciences, University of Exeter, UK, 2008.

Broughton, T. "The Bhopal Disaster and Its Aftermath: A Review." *Environmental Health* 4, no. 6 (2005): 1–6.

Brown, Lester. *Eco-Economy: Building and Economy for the Earth*. New York, NY: W. W. Norton & Co, 2001.

Brown, P. and E .J. Mikkelsen. *No Safe Place: Toxic Waste, Leukemia and Community Action*. Berkeley: University of California Press, 1990.

Brown, Phil. "Popular Epidemiology and Toxic Waste Contamination: Lay and Professional Ways of Knowing." *Journal of Health and Social Behavior* 33 (September 1992): 267–281.

————. *Toxic Exposures: Contested Illnesses and the Environmental Health Movement*. New York, NY: Columbia University Press, 2007.

Bryant, Bunyan and Paul Mohai, eds. *Race and the Incidence of Environmental Hazards: A Time for Discourse*. Boulder, CO: Westview Press, 1992.

Bullard, Robert D. *Dumping in Dixie: Race, Class, and Environmental Quality*. Boulder, CO: Westview Press, 1990, 2000.

————. "Anatomy of Environmental Racism and the Environmental Justice Movement." pp. 15–23 in *Confronting Environmental Racism: Voices from the Grassroots*, edited by Robert d. Bullard. Boston, MA: South End Press, 1993.

————. "Environmental Justice in the Twenty First Century." pp. 19–42 in *The Quest for Environmental Justice: Human Rights and the Politics of Pollution*, edited by Robert D. Bullard. San Francisco, CA: Sierra Club Books, 2005.

Bullard, Robert D. "Differential Vulnerabilities: Environmental and Economic Inequality and Government Response to Unnatural Disasters." *Social Research* 75, no. 3 (2008): 753–784.

Bullard, R. D., P. Mohai, R. Saha, and B. Wright. *Toxic Wastes and Race at Twenty, 1987–2007: A Report Prepared for the United Church of Christ Justice & Witness Ministries.* Cleveland, OH: United Church of Christ, 2007.

Buttel, Frederick H. and W. L. Flinn. "The Structure of Support for the Environmental Movement, 1968–1970." *Rural Sociology* 39 (Spring 1974): 56–59.

Cantanhede, Alvaro. *Hazardous Waste Characterization and Classification Summary. Pan American Center for Sanitary Engineering and Environmental Sciences (CEPIS).* Lima, Peru: CEPIS, 1994.

Carroll, Chris. "High-Tech Trash: Will Your Discarded TV or Computer End Up in Ghana?" *National Geographic* 213, no. 1 (2008): 64–81.

Carruthers, David V. "The Globalization of Environmental Justice: Lessons from the US-Mexico Border." *Society and Natural Resources* 21 (2008): 556–568.

Carson, Rachel. *Silent Spring.* Boston, MA: Houghton Mifflin Co, 1962.

Catton, W. R. and R. E. Dunlap. "A New Ecological Paradigm for Post-exuberant Sociology." *American Behavioral Scientist* 24 (September/October 1980): 15–47.

Center for Disease Control (CDC). *Preventing Lead Poisoning in Young Children.* Atlanta, GA: CDC, 1991.

Chapman, Stephen R. *Environmental Law and Policy.* Upper Saddle River, NJ: Prentice Hall, 1998.

Chedrese, J. P., M. Piasek, and M. C. Henson. "Cadmium as an Endocrine Disruptor in the Reproductive System." *Immunology, Endocrine and Metabolic Agents in Medicinal Chemistry* 6, no. 1 (2006): 27–35.

Clapp, Jennifer. *Toxic Exports: The Transfer of Hazardous Wastes from Rich to Poor Countries.* Ithaca, NY: Cornell University Press, 2001.

Clapp, R. W. "Mortality among US Employees of a Large Computer Manufacturing Company: 1969–2001." *Environmental Health* 5, no. 30 (2006): 1–10.

Clapp, R. "The Love Canal Story Is Not Finished." *Environmental Health Perspectives* 117, no. 2 (2009): A54.

Cohen, G. and J. O'Connor, eds. *Fighting Toxics: A Manual for Protecting Your Family, Community, and Workplace.* Washington, DC: Island Press, 1990.

Colborn, Theo, Dianne Dumanoski, and John Peterson Myers. *Our Stolen Future: Are We Threatening Our Fertility, Intelligence, and Survival?—A Scientific Detective Story.* New York: Penguin, 1996.

Cole, Luke W. and Sheila R. Foster. *From the Ground Up: Environmental Racism and the Rise of the Environmental Justice Movement.* New York: New York University Press, 2001.

Comfort, Susan. "Struggle in Ogoniland: Ken Saro-Wiwa and the Cultural Politics of Environmental Justice." pp. 229–246 in *Environmental Justice Reader: Politics, Poetics, and Pedagogy,* edited by J. Adamson, M. M. Evans, and R. Stein. Tucson: University of Arizona Press, 2002.

Comte, A. and A. Flüry-Herard. "Radioactive Waste: What Health Effects, or Risks?" *CLEFS CEA* 53 (Winter 2005–2006): 9–11.

Consonni, D., A. C. Pesatori, C. Zocchetti, R. Sindaco, L. C. D'Oro, M. Rubagotti, and P. A. Bertazzi. "Mortality in a Population Exposed to Dioxin after the Seveso, Italy, Accident in 1976: 25 Years Follow-Up." *American Journal of Epidemiology* (2008): 1–12. http://aje.oxford journals.org (accessed March 29, 2010).

Cotgrove, S. *Catastrophe or Cornucopia: The Environment, Politics, and the Future.* New York: John Wiley and Sons, 1982.

Crawford, M. *Toxic Waste Sites: An Encyclopedia of Endangered America.* Santa Barbara, CA: ABC-CLIO, Inc., 1997.

Crinnion, Walter J. "Environmental Medicine Part 1: The Human Burden of Environmental Toxins and Their Common Health Effects." *Alternative Medicine Review* 5, no. 1 (2000): 52–63.

Cunningham, William P. and Barbara W. Saigo, *Environmental Science: A Global Concern.* New York, NY: McGraw-Hill, 1999.

Cunningham, William P., Mary A. Cunningham, and Barbara W. Saigo. *Environmental Science: A Global Concern.* New York, NY: McGraw-Hill, 2007.

Cuthberson, Beverly H. and Joanne M. Nigg. "Technological Disaster and the Nontherapeutic Community: A Question of True Victimization." *Environment and Behavior* 19 (1987): 462–483.

Davidar, D. "Beyond Bhopal: The Toxic Waste Hazard in India." *Ambio* 14, no. 2 (1985): 112–116.

Dayaneni, G. and J. Doucette. *System Error: Toxic Tech Poisoning People and Planet.* San Jose, CA: Silicon Valley Toxics Coalition (SVTC), 2005.

DeMarchi, B., S. Funtowicz, and J. Ravetz. "Seveso: A Paradoxical Classic Disaster." Chapter 4 in *Long Road to Recovery: The Community Responses to Industrial Disaster,* edited by J .K. Mitchell. New York, NY: United Nations University Press, 1996.

Department of the Environment and Heritage. Department of the Environment and Heritage Annual Report Volume 1 & 2, 2004–2005. http://www.deh.gov.au/about/publications/annual-report (accessed December 23, 2009).

Dhara, V. R., R. Dhara, S. D. Acquilla, and P. Cullinan. "Personal Exposure and Long-Term Health Effects in Survivors of the Union Carbide Disaster at Bhopal." *Environmental Health Perspectives* 110, no. 5 (2002): 487–500.

Dhara, V. R. and R. Dhara. "The Union Carbide Disaster in Bhopal: A Review of Health Effects." *Archives of Environmental Health* 57, no. 5 (2002): 391–404.

Dhara, V. R. and D. Kriebel. "An Exposure-Response Method for Assessing the Long-Term Health Effects of the Bhopal Gas Disaster." *Disasters* 17, no. 4 (1993): 281–290.

Dinham, B. and S. Surangi. "The Bhopal Gas Tragedy 1984 to the Evasion of Corporate Responsibility." *Environment and Urbanization* 14, no. 1 (2002): 89–99.

Douglas, M. and A.Wildavsky. *Risk and Culture*. Berkeley, CA: University of California Press, 1982.

Drabek, Thomas E. *Human System Responses to Disaster: An Inventory of Sociological Findings*. New York: Springer-Verlag, 1986.

Drotman, Peter D. (1985). "Chemicals, Health, and the Environment." pp. 47–77 in *Introduction to Environmental Health*, edited by D. S. Blumenthal. New York: Springer Publishing Company, 1985.

D'Silva, Themistocles. *The Black Box of Bhopal: A Closer Look at the World's Deadliest Industrial Disaster*. Oxford, UK: Trafford Publishing Co., 2006.

Dunlap, R. E. and A. G. Mertig. "The Evolution of the US Environmental Movement from 1970 to 1990: An Overview." pp. 1–10 in *American Environmentalism: The US Environmental Movement, 1970–1990*, edited by R. E. Dunlap and A. G. Mertig. New York: Taylor and Francis, 1992.

Dunlap, R. E., G. H. Gallup, and A. M. Gallup. "Of Global Concern: Results of the Health of the Planet Survey." *Environment* 35, no. 9 (1993): 7–39.

Eckley, Noelle. "Traveling Toxics: The Science, Policy, and Management of Persistent Organic Pollutants." *Environment* 43, no. 7 (2001): 23–36.

Edelstein, Michael R. *Contaminated Communities: The Social and Psychological Impacts of Residential Toxic Exposure*. Boulder, Co: Westview Press, 1988.

Edelstein, Michael R. and William J. Makofske. *Radon's Deadly Daughters: Social Science, Environmental Policy, and the Politics of Risk*. Landham, MD: Rowman and Littlefield Publishers, 1998.

Edwards, T. "Contamination of Community Water Sources in Bhopal, India." 2005. http://bhopal.org/fid/admin/content/documents/contaminationbrief.pdf (accessed November 5, 2009).

Engler, R. "Many Bhopals: Technology Out of Control." *The Nation* 240, no. 16 (1985): 488–500.

Epstein, S. S., L. O. Brown, and C. Pope. *Hazardous Waste in America.* San Francisco, CA: Sierra Club Books, 1982.

Erikson, K. A. *New Species of Trouble: Explorations in Disaster, Trauma, and Community.* New York, NY: W. W. Norton, 1994.

————.*Everything in Its Path: Destruction of Community in the Buffalo Creek Flood.* New York: Simon and Schuster, 1976.

Et Forecasts. *World PC Market.* 2009. http://www.etforecasts.com /products/ES-PC WW1203.htm (accessed December 30, 2009).

Faber, D. *Capitalizing on Environmental Injustice: The Polluter-Industrial Complex in the Age of Globalization.* Lanham, MD: Rowman and Littlefield Publishers, 2008.

Faupel, C. E., C. Bailey, and G. Griffin. "Local Media Roles in Defining Hazardous Waste as a Social Problem: The Case of Sumter County, Alabama." *Sociological Spectrum* 11 (1991): 293–319.

Field, Barry C. and Martha K Field. *Environmental Economics: An Introduction.* New York, NY: McGraw-Hill/Irwin, 2002.

Figueroa, R. "Teaching for Transformation: Lessons from Environmental Justice." pp. 311–330 in *The Environmental Justice Reader: Politics, Poetics, and Pedagogy*, edited by J. Adamson, M. M. Evans, and R. Stein. Tucson: University of Arizona Press, 2002.

Finch, S. 1994. "Poisoned Property." *The Times Picayune*, August 21, 1994, A4.

Finlay, Alan. "E-waste Challenges in Developing Countries: South Africa Case Study." APC Issue Papers. Association for Progressive Communications, November, 2005. http://www.apc.org/en/system/ files/e-waste_EN.pdf (accessed December 23, 2009).

Finucane, Melissa L., Paul. Slovic, C. K. Mertz, J. Flynn, and T .A. Satterfield. "Gender, Race, and Perceived Risk: The White Male Effect." *Health, Risk, and Society* 2, no. 2 (2000): 161–172.

Fitzpatrick, Kevin and Mark LaGory. *Unhealthy Places: The Ecology of Risk in the Urban Landscape.* New York, NY: Routledge, 2000.

Flynn, J., P. Slovic, and C. K. Mertz. "Gender, Race, and Perception of Environmental Health Risks." *Risk Analysis*, 14 (1994): 1101–1108.

Frank, D. J., A. Hironaka, and E. Schofer. "The Nation-State and the Natural Environment over the Twentieth Century." *American Sociological Review* 65, no. 1 (2000): 96–116.

Fransson, N. and T. Garling. "Environmental Concern: Conceptual Definitions, Measurement Methods, and Research Findings." *Environmental Psychology* 19, no. 4 (1999): 369–382.

Fredholm, L. "Chemical Testing: Sweden to Get Tougher on Lingering Compounds." *Science* 290, no. 5497 (2000): 1663–1666.

Freudenberg, N. *Not in Our Backyards: Community Action for Health and the Environment.* New York, NY: Monthly Review Press, 1984.

Freudenburg, William R. "Contamination, Corrosion, and Social Order: An Overview." *Current Sociology* 45, no. 3 (1997): 19–40.

———. "Risk and Recreancy: Weber, the Division of Labor, and the Rationality of Risk Perceptions." *Social Forces* 71, no. 4 (1993): 909–932.

Frey, S. R. "The International Traffic in Hazardous Wastes." *Journal of Environmental Systems* 23, no. 2 (1994–95): 165–177.

———. "The Transfer of Core-Based Hazardous Production Processes to the Export Processing Zones of the Periphery: The Maquiladora Centers of Northern Mexico." *Journal of World Systems Research*, IX, no. 2 (2003): 317–354.

Fritz, C. E. and H. B. Williams. "The Human Being in Disasters: A Research Perspective." *Annals of the American Academy of Political and Social Science* 309 (1957): 42–57.

Funabashi, H. "Minamata Disease and Environmental Governance." *International Journal of Japanese Sociology* 15, no. 1 (2006): 7–25.

Gehlawat, J. K. "Bhopal Disaster—A Personal Experience." Unpublished paper, 2005. http://www.utk.ac.in/che/jpg/papersb/full%20 papers/G%20-%2013.doc (accessed November 28, 2009).

Gensburg, L. J., C. Pantea, E. Fitzgerald, A. Stark, S. Hwang, and N. Kim. "Mortality among Former Love Canal Residents." *Environmental Health Perspectives* 117, no. 2 (2009): 209–216.

Gerrard, Michael B. *Whose Backyard, Whose Risk: Fear and Fairness in Toxic and Nuclear Waste Siting.* Cambridge: MIT Press, 1995.

Giddens, Anthony. The *Consequences of Modernity.* Cambridge: Polity Press, 1990.

Gill, Duane A. "Secondary Trauma or Secondary Disaster? Insights from Hurricane Katrina." *Sociological Spectrum* 27, no. 6 (2007): 613–632.

Gill, Duane A. and J. Steven Picou. "Technological Disaster and Chronic Community Stress." *Society and Natural Resources* 11, no. 8 (1998): 795–815.

Girdner, Eddie J. and Jack Smith. *Killing Me Softly: Toxic Waste, Corporate Profit, and the Struggle for Environmental Justice.* New York: Monthly Review Press, 2002.

Goklany, Indur M. *The Precautionary Principle: A Critical Appraisal of Environmental Risk Assessment.* Washington, DC: CATO Institute, 2001.

Gould, Kenneth A., David N. Pellow, and Allan Schnaiberg. *The Treadmill of Production: Injustice and Sustainability in the Global Economy.* Boulder, CO: Paradigm Publishers, 2008.

Government Accounting Office (GAO). *Siting of Hazardous Waste Landfills and Their Correlation with Racial and Economic Status of Surrounding Communities.* Washington, DC: GAO, 1983.

————. "Environmental Justice: EPA Should Devote More Attention to Environmental Justice When Developing Clean Air Rules." Report to the Ranking Member, Sub-Committee on Environment and Hazardous Materials, Committee on Energy and Commerce, House of Representatives. 2005. http://www.gao.gov/new.items/d05 289. pdf (accessed August 21, 2010).

————. "Electronic Waste: EPA Needs to Better Control Harmful U.S. Exports Through Stronger Enforcement and More Comprehensive Regulation." Report to the Chairman, Committee on Foreign Affairs, House of Representatives, Washington, DC: GAD-08-1044. 2008. http://www.gao.gov/new.items/do647.pdf (accessed February 28, 2010).

————. *International Trade: Mexico's Maquiladora Decline Affects US-Mexico Border Communities and Trade; Recovery Depends in Part on Mexico's Actions.* Washington, DC: GOA, 2003.

Graham, Mary and Catherine Miller. "Disclosure of Toxic Releases in the United States." *Environment* 43, no. 8 (2001): 9–20.

Greenpeace. *The Database of Known Hazardous Waste Exports from OECD to Non-OECD Countries, 1989–1994.* Washington, DC: Greenpeace, 1994.

Greider, T. and L. Garkovich. "Landscapes: The Social Construction of Nature and the Environment." *Rural Sociology* 59 (1994): 1–24.

Grineski, Sara E., T. W. Collins, M. L. R. Aguilar, and R. Aldouri. "No Safe Place: Environmental Hazards and Injustice along Mexico's Northern Border." *Social Forces* 88, no. 5 (2010): 2241–2266.

Grineski, Sara E. and T. W. Collins. "Exploring Patterns of Environmental Injustice in the Global South: Maquiladoras in Ciudad Juárez, Mexico." *Population and Environment* 29, no. 6 (2008): 247–270.

Grossman, Elizabeth. *High-Tech Trash: Digital Devices, Hidden Toxics, and Human Health.* Washington, DC: Island Press/Shearwater Books, 2006.

Grossman, G. M. and H. R. Portter. "A Trend Analysis of Competing Models of Environmental Attitudes." Working Paper no. 127. West Lafayette, IN: Department of Sociology and Anthropology, Purdue University, 1977.

Guber, D. L. "Environmental Concern and the Dimensionality Problem: A New Approach to an Old Predicament." *Social Science Quarterly* 77, no. 3 (1996): 644–662.

Guillette, Elizabeth, Maria Mercedes Meza, Maria Guadalupe Aquilar, Alma Delia Soto, and Idalia Enedina Garcia. "An Anthropological Approach to the Evaluation of Preschool Children Exposed to Pesticides in Mexico." *Environmental Health Perspectives* 106, no. 6 (1998): 347–353.

Hackett, Paul. *Conservation and the Consumer: Understanding Environmental Concern.* New York: Routledge, 1995.

Hallman, W. K. and A. Wandersman. "Attribution of Responsibility and Individual and Collective Coping with Environmental Threats." *Journal of Social Issues* 48, no. 4 (1992): 101–118.

Hamada, R. and M. Osame. "Minamata Disease and Other Mercury Syndromes." pp. 337–351 in *Toxicology of Metals*, edited by L. W. Chang. New York: CRC Lewis Publishers, 1996.

Hamilton, L. C. "Concern about Toxic Wastes: Three Demographic Predictors." *Sociological Perspectives* 28, no. 4 (1985): 463–486.

Hannigan, John A. *Environmental Sociology: A Social Constructionist Perspective.* New York, NY: Routledge, 1995.

Hauser, Russ, Larisa Altshul, Zuying Chen, Louise Ryan, James Overstreet, Issac Schiff, and David Christiani. "Environmental Organochlorines and Semen Quality: Results of a Pilot Study." *Environmental Health Perspectives* 110, no. 3 (2002): 229–233.

Hernan, R. E. *This Borrowed Earth: Lessons from the 15 Worst Environmental Disasters around the World.* New York, NY: Palgrave MacMillan, 2010.

Hofrichter, R. "Introduction: Critical Perspectives on Human Health and the Environment." pp. 1–15 in *Reclaiming the Environmental Debate: The Politics of Health in a Toxic Culture*, edited by R. Hofrichter. Cambridge: MIT Press, 2000.

Homberger, E., G. Reggiano, J. Sambeth, and H. K. Wipf. "The Seveso Accident: Its Nature, Extent and Consequences." *Annals of Occupational Hygiene* 22, no. 4 (1979): 327–370.

Hooks, G. and C. L. Smith. "The Treadmill of Destruction: National Sacrifice Areas and Native Americans." *American Sociological Review* 69 (2004): 558–575.

Höpfl, H. and S. Matilal. "Complexity and Catastrophe: Disentangling the Complex Narratives of the Union Carbide Disaster in Bhopal." *Emergent: Complexity and Organization* 7, no. 3–4 (2005): 64–73.

Huo, X., L. Peng, X. Xu, L. Zheng, B. Qiu, Z. Qi, B. Zhang, D. Han, and Z. Piao. "Elevated Blood Lead Levels of Children in Guiyu, an

Electronic Waste Recycling Town in China." *Environment Health Perspectives* 115, no. 7 (2007): 1113–1117.

Huberty, C .J. *Applied Discriminant Analysis.* New York: John Wiley and Sons, 1994.

Humphrey, Craig R., Tammy L. Lewis, and Frederick H Buttel. *Environment, Energy, and Society: A New Synthesis.* Belmont, CA: Wadsworth/Thomson Learning, 2002.

Hunter, L. M. "A Comparison of the Environmental Attitudes, Concern, and Behaviors of Native-Born and Foreign-Born US Residents." *Population and Environment* 21, no. 6 (2000): 565–580.

Ihonvbere, J. O. "The State and Environmental Degradation in Nigeria: A Study of the 1988 Toxic Waste Dump in Koko." *Journal of Environmental Systems* 23, no. 3 (1994–1995): 207–227.

Iles, A. "Mapping Environmental Justice in Technology Flows: Computer Waste Impacts in Asia." *Global Environmental Politics* 4, no. 4 (2004): 76–107.

Inglehart, Ronald. *Culture Shift in Advanced Industrial Society.* Princeton, NJ: Princeton University Press, 1990.

International Federation of Red Cross and Red Crescent Societies. *World Disaster Report: Focus on Discrimination.* Bloomfield, CT: Kumarian Press, 2007.

Isaac, L., E. Mutran, and S. Stryker. "Political Protest Orientations among Black and White Adults." *American Sociological Review* 45, no. 2 (1980): 191–213.

ITU. Information Society Statistical Profiles: Africa. Geneva: ITU, 2009a.

ITU. Information Society Statistical Profiles: Americas. Geneva: ITU.

Jablonski, Susan M. "Radioactive Waste." 2009. http://www.pollutionis-sues.com/Pl-Re/Radiocactive-Waste.html (accessed June 12, 2009).

Jain, R. K., L. V. Urban, and H. E. Balbach. *Environmental Assessment.* New York, NY: McGraw-Hill, 1993.

Jasanoff, S. "Symposium. The Bhopal Disaster Approaches 25: Looking Back to Forward—Bhopal's Trials of Knowledge and Ignorance." *New England Law Review* 42 (Summer 2008): Rev. 679.

Johnson, G. S. "Environmental Justice: A Brief History and Overview." pp. 17–45 in *Environmental Justice in the New Millennium: Global Perspectives on Race, Ethnicity, and Human Rights,* edited by Filomina C. Steady. New York: Palgrave MacMillan, 2009.

Jones, K. C. and P. deVoogt. "Persistent Organic Pollutants (POPs): State of the Science." *Environmental Pollution* 100 (1999): 209–221.

Jones, R. E. "Black Concern for the Environment: Myth versus Reality." *Society and Natural Resources* 11 (1998): 209–228.

Jones, R. E. and L. F. Carter. "Concern for the Environment among Black Americans: An Assessment of Common Assumptions." *Social Science Quarterly* 75 (1994): 560–579.

Jones, R. E. and R. E. Dunlap. "The Social Bases of Environmental Concerns: Have They Changed Over Time?" *Rural Sociology* 57 (1992): 28–47.

Kahhat, R. and E. Williams. "Product or Waste? Importation and End-of-Life Processing of Computers in Peru." *Environmental Science and Technology* 43, no. 5 (2009): 6010–6016.

Kalof, L., T. Dietz, G. Guagnano, and P. C. Stern. "Race, Gender and Environmentalism: The Atypical Values and Beliefs of White Men." *Race, Gender & Class* 9, no. 2 (2002): 1–19.

Kapoor, R. "The Psychological Consequences of an Environmental Disaster: Selected Case Studies of the Bhopal Gas Tragedy." *Population and Environment* 13, no. 3 (1992): 209–215.

Karan, P. P., W. A. Bladen, and J. R. Wilson. "Technological Hazards in the Third World." *The Geographical Review* 76, no. 2 (1986): 195–208.

Kasuya, M. "Recent Epidemiological Studies on Itai-Itai Disease as a Chronic Cadmium Poisoning in Japan." *Water, Science and Technology* 42, no. 7–8 (2000): 147–155.

Kempton, W., J. S. Boster, and J. A. Hartley. *Environmental Values in American Culture.* Cambridge: MIT, 1995.

Kennedy, D. D. "Death and Justice: Environmental Tragedy and the Limits of Science." 1999. http://www2.shore.net/~dkennedy/woburn_trial.html (accessed November 5, 2009).

———. "Stalking Woburn's Mystery Killer." 1997. http://www2.shore.net/~dkennedy/woburn_mit.html (accessed November 20, 2009).

Klineberg, Stephen L., M. McKeever, and B. Rothenbach. "Demographic Predictors of Environmental Concern: It Does Make a Difference How It's Measured." *Social Science Quarterly* 79, no. 4 (1998): 734–753.

Kowaleski, D. and K. L. Porter. "Environmental Concern among Local Citizens: A Test of Competing Perspectives." *Journal of Political and Military Sociology* 21 (Summer 1993): 37–62.

Koziol, Anna and Janusz Pudykiewicz. "Global-Scale Environmental Transport of Persistent Organic Pollutants." *Chemosphere* 45 (2001): 1181–1200.

Krause, D. "Environmental Consciousness: An Empirical Study." *Environment and Behavior* 25, no. 1 (1993): 126–142.

Kreps, G. A. "Disasters and the Social Order." *Sociological Theory* 3 (1985): 49–65.

Kreps, G. A. and G. A. Drabek. "Disasters as Nonroutine Social Problems." *International Journal of Mass Emergencies and Disasters* 14 (1996): 129–153.

Krueger, Jonathan. "The Basel Convention and the International Trade in Hazardous Wastes." pp. 43–51 in *Olav Yearbook of International Cooperation on Environment and Development*, edited by Schram Stokke and Oystein B. Thommessen. London: Earthscan, 2001.

Kubasek, Nancy K. and Gary S. Silverman. *Environmental Law*. Upper Saddle, NJ: Prentice Hall, 2000.

———. *Environmental Law*. Upper Saddle, NJ: Pearson/Prentice Hall, 2005.

Kuehn, Robert R. "A Taxonomy of Environmental Justice." *Environmental Law Reporter*, 30, no. 9 (2000): 10681–10703.

Kuehr, Ruediger, German T. Velasquez, and Eric Williams. "Computers and the Environment: An Introduction to Understanding and Managing Their Impacts." pp. 1–15 in *Computers and the Environment: Understanding and Managing Their Impacts*, edited by RuedigerKuehr and Eric Williams. Norwell, MA: Kluwer Academic Publishers, 2003.

Kummer, K. "The Basel Convention: Ten Years On." *RECIEL* 7, no. 3 (1998): 227–236.

Labunska, I., A. Stephenson, K. Brigden, R. Stringer, D. Santillo, and P. A. Johnston. *The Bhopal Legacy*. Exeter, UK: Greenpeace, 1999.

LaGrega, Michael D., P. L. Buckingham, J. C. Evans, and Environmental Resources Management (ERM). *Hazardous Waste Management*. New York, NY: McGraw-Hill, 2001.

Lallas, Peter. "The Stockholm Convention of Persistent Organic Pollutants." *American Journal of International Law* 95 (2002): 692–708.

———. "The Role of Process and Participation in the Development of Effective International Environmental Agreements: A Study of the Global Treaty on Persistent Organic Pollutants." *Journal of Environmental Law* 19 (2000/2001): 83–153.

LaPierre, D. and Moro, J. *Five Past Midnight in Bhopal*. London, UK: Scribner, 2002.

Lave, L. B. and A. C. Upton, eds. *Toxic Chemicals, Health, and the Environment*. Baltimore, MD: Johns Hopkins University Press, 1987.

Lavell, M. and M. Coyle. "Unequal Protection: Racial Divide in Environmental Law." *National Law Journal* 15 (Supplement September 1992): 52–54.

Lemons, H. "Physical Characteristics of Disasters: Historical and Statistical Review." *Annals of the American Academy of Political and Social Science* 309 (1957): 1–14.

Lester, J. P., D. W. Allen, and K. M. Hill. *Environmental Injustice in the United States: Myths and Realities.* Boulder, CO: Westview Press, 2001.

Leung, A. O. W., N. S. Duzgorem-Aydin, K. C. Cheung, and M. H. Wong. "Heavy Metals Concentrations of Surface Dust from e-Waste Recycling and Its Human Health Implications in Southeast China." *Environmental Science and Technology* 42, no. 7 (2008): 2674–2680.

Leung, A., Z. W. Cai, and M. H. Wong. "Environmental Contamination from Electronic Waste Recycling at Guiyu, Southeast China." *Journal of Material Cycles Waste Management* 8 (2006): 21–33.

Levi, D., S. Kocher, and R. Aboud. "Technological Disasters in Natural and Built Environments." *Environment and Behavio*r 33, no. 1 (2001): 78–92.

Levine, A. G. *Love Canal: Science, Politics and People.* Lexington, MA: Lexington Books, 1982.

Lichtveld, M. Y. and B. L. Johnson. "Public Health Implications of Hazardous Waste Sites in the United States." Paper Presented at the ATSDR Hazardous Waste Conference, 1993. http://www.atsdr.cdc.gov /cxic.html (accessed February 10, 2009).

Lifton, R. J. "Nuclear Energy and the Wisdoms of the Body." *Bulletin of the Atomic Scientist* (September 1976): 16–20.

Lipman, Z. "A Dirty Dilemma: The Hazardous Waste Trade." *Harvard International Review* 23, no. 4 (2002): 67–71.

Litmanan, T. "Environmental Conflict as a Social Construction: Nuclear Waste Conflicts in Finland." *Society and Natural Resources* 9 (1996): 523–535.

Luhmann, N. *Risk: A Sociological Theory.* New York, NY: Walter de Gruyter, 1993.

Lupton, D. *Risk.* New York: Routledge, 1999.

Luther, L. "Managing Electronic Waste: Issues with Exporting E-Waste." *Congressional Research Service Report for Congress #7-5700.* 2009. http://www.fas.org/sgp/crs/misc/R40850.pdf (accessed December 29, 2009).

Ma, J., R. Addink, S. Yun, J. Cheng, W. Wang, and K, Kannan. "Polybrominated Dibenzo-p-dioxins/Dibenzofurans and Polybrominated Diphenyl Ethers in Social, Vegetation, Workshop-Floor Dust, and Electronic Shredder Residue from an Electronic Waste Recycling Facility and Soils from a Chemical Industrial

Complex in East China." *Environmental Science and Technology* 43, no. 19 (2009): 7350–7356.

Martinez-Alier, J. *The Environmentalism of the Poor: A Study of Ecological Conflicts and Valuation.* Cheltenham, UK: Edward Elgar Publishing Ltd, 2002.

Maslow, A. H. *Motivation and Personality.* New York, NY: Harper and Row Publishers, 1970.

Matsuura, Nobuo, Tomoaki Uchiyama, Hiroshi Tada, Yosikazu Nakamura, Naomi Kondo, Matastoshi Morita, and Masaru Fukushi. "Effects of Dioxins and Polychlorinated Biphenyls (PCBs) on Thyroid Function in Infants Born in Japan: The Second Report from Research on Environmental Health." *Chemosphere* 45 (2001): 1167–1171.

Maxwell, N. I. *Understanding Environmental Health: How We Live in the World.* Sudbury, MA: Jones and Bartlett Publishers, 2009.

McGinn, Anne P. "Phasing Out Persistent Organic Pollutants." pp. 79–100 in *State of the World,* edited by Brown L. R. et al. New York, NY: W. W. Norton, 2000.

———. "Reducing Our Toxic Burden." pp. 75–100 in *State of the World,* edited by Linda Starke. New York: Norton & Company, 2002.

McGurty, E. *Transforming Environmentalism: Warren County, PCBs, and the Origins of Environmental Justice.* New Brunswick, NJ: Rutgers University Press, 2009.

McKenna, Alan. "Computer Waste: A Forgotten and Hidden Side to the Global Information Society." *Environmental Law Review* 9, no. 2 (2007): 116–131.

McLachlan, G. J. *Discriminant Analysis and Statistical Pattern Recognition.* New York: John Wiley and Sons, 1992.

Medvedev, Zhores. *The Legacy of Chernobyl.* New York, NY: W. W. Norton, 1990.

Mertig, Angela G. and Riley E. Dunlap. "Environmentalism, New Social Movements, and the New Class: A Cross-National Investigation." *Rural Sociology* 66, no. 1 (2001): 113–136.

Milbrath, L. *Environmentalists: Vanguard for a New Society.* Albany: State University of New York Press, 1984.

Miller, E. W. and R. M. Miller. *Environmental Hazards: Toxic Waste and Hazardous Materials. A Reference Handbook.* Denver, CO: ABC-CLIO, Inc., 1991.

Miller, G. T. Jr. *Environmental Science.* Belmont, CA: Wadsworth, 2001.

Mitchell, J. K. "Signposts on the Road to Recovery." Chapter 9 in *Long Road to Recovery: The Community Responses to Industrial Disasters,* edited by J. K. Mitchell. New York: United Nations University Press, 1996.

Mitchell, R. "Silent Spring/Solid Majorities." *Public Opinion* 2 (1979): 16–20, 55.

Mohai, Paul. "Black Environmentalism." *Social Science Quarterly* 71 (1990): 744–765.

———. "Dispelling Old Myths: African American Concern for the Environment." *Environment* 45, no. 5 (2003): 11–26.

Mohai, Paul and Bunyan Bryant. "Environmental Racism: Reviewing the Evidence." pp. 163–176 in *Race and the Incidence of Environmental Hazards: A Time for Discourse*, edited by B. Bryant and Paul Mohai. Boulder, CO: Westview Press, 1992.

Mohai, Paul and Bunyan Bryant. "'Is There a 'Race Effect' on Concern for Environmental Quality?" *Public Opinion Quarterly* 62 (1998): 475–505.

Mohai, Paul and B. W. Twight. "Age and Environmentalism: An Elaboration of Buttel Model Using National Survey Evidence." *Social Science Quarterly* 68, no. 4 (1987): 798–815.

Mohai, Paul and R. Saha. "Racial Inequality in the Distribution of Hazardous Waste: National Level Reassessment." *Social Problems* 54, no. 3 (2007): 343–370.

Morehouse, W. and M. A. Subramaniam. *The Bhopal Tragedy: What Really Happened and What It Means for American Workers and Communities at Risk: A Report for the Citizens Commission on Bhopal*. New York: Council on International and Public Affairs (CIPA), 1986.

Morrison, D. E. "How and Why Environmental Consciousness Has Trickled Down." pp. 187–122 in *Distribution Conflict in Environmental Resource Policy*, edited by A. Schnaiberg, N. Watts, and K. Zimmerman. New York: St. Martin's Press, 1986.

Morrison, D. E. and R. E. Dunlap. "Environmentalism and Elitism: A Conceptual and Empirical Analysis." *Environmental Management* 10 (1986): 581–589.

Moser, Andreas G. and Michael S. McLachlan. "The Influence of Dietary Concentration on the Absorption and Excretion of Persistent Lipophilic Organic Pollutants in the Human Intestinal Tract." *Chemosphere* 45, no. 2 (2001): 201–211.

Moyers, B. *Global Dumping Ground: The International Traffic in Hazardous Waste*. Washington, DC: Seven Locks Press, 1990.

Myers, J. P. "The Latest Hormone Science, Part 4: Disrupting Life's Messages." *Rachel's Environment and Health News* 753 (October 3, 2002): 1–5. http://www.racchel.org

Narayan, T. "Health Impact of Bhopal Disaster: An Epidemiological Perspective." *Economic and Political Weekly* 25, no. 34 (1990): 1905–1914.

Nawrot, T. S., E. V. Hecke., L. Thijs, T. Richart, T. Kuzentsova, Y. Jin, J. Vomgronsveld, H. A. Roels, and J. A. Staessen. "Cadmium-Related Mortality and Long-Term Secular Trends in the Cadmium Body Burden of an Environmentally Exposed Population." *Environmental Health Perspectives* 16, no. 2 (2008): 1620–1628, *U.S.G.S. Mineral Commodity Summaries: Cadmium.* http://minerals.usgs.gov/minerals/pubs/commodity/cadmium/MCS_2008_Cadmium.pdf (accessed December 7, 2009).

Nebel, B. J. and R. T. Wright. *Environmental Science: The Way the World Works.* Upper Saddle River, NJ: Prentice Hall, 2000.

New World Encyclopedia."Electronic Waste." 2009. http://www.newworldencyclopedia.org/entry/Electronic Waste (accessed December 29, 2009).

Nogawa, K. and T. Kido. "Itai-Itai Disease and Health Effects of Cadmium." pp. 353–369 in *Toxicology of Metals*, edited by L. W. Chang. New York: CRC-Lewis Publishers, 1996.

Novtny, Patrick. *Where We Live, Work and Play: The Environmental Justice Movement and the Struggle for a New Environmentalism.* Westport, CT: Praeger, 2000.

Nriagu, J. O. "History of Global Metal Production." *Science* 22 (April 12, 1996): 223–224.

O'Connor, J."The Toxic Crisis." pp. 11–24 in *Fighting Toxics: A Manual for Protecting Your Family, Community, and Workplace*, edited by G. Cohen and J. O'Connor. Washington, DC: Island Press, 1990.

Okonta, Ike and Oronto Douglas. *Where the Vultures Feast: Shell, Human Rights, and Oil in the Niger Delta.* San Francisco, CA: Sierra Club Books, 2001.

Oliver-Smith, A. "Anthropological Research on Hazards and Disasters." *Annual Review of Anthropology* 25 (1996): 303–328.

Olsen, M., D. Lodwick, and R. Dunlap. *Viewing the World Ecologically.* Boulder, CO: Westview Press, 1992.

Olsson, I. M., I. Bensrd, T. Lundh, H. Ottoson, S. Skerfving, and A. Oskarsson. "Cadmium in Blood and Urine—Impact of Sex, Age, Dietary Intake, Iron Status, and Former Smoking—Association of Renal Effects." *Environmental Health Perspectives* 110 (2002): 1185–1190.

Ostheimer, J. M. and L. G. Ritt. *Environment, Energy, and Black Americans.* Beverly Hills, CA: Sage Publications, 1976.

Parker, J. D. and M. H. McDonough. "Environmentalism of African Americans: An Analysis of the Subculture and Barriers Theories." *Environment and Behavior* 31 (1999): 155–177.

Pedersen, O. W. "Environmental Principles and Environmental Justice." *Environmental Law Review* 12 (2010): 26–49.

Pellow, D. N. *Resisting Global Toxics: Transnational Movements for Environmental Justice.* Cambridge: MIT Press, 2007.

Perrow, C. *Normal Accidents: Living with High Risk Technologies.* New York: Basic Books, 1984.

Pesatori, A. C., D. Consonni, M. Rubagotti, P. Grillo, and A. Bertazzi. "Cancer Incidence in the Population Exposed to Dioxin after the 'Seveso Accident': Twenty Years of Follow-Up." *Environmental Health* 8, no. 39 (2009): 1–11.

Phillips, A. S., Y. Hung, and P. A. Bosela. Love Canal Tragedy." *Journal of Performance of Constructed Facilities* 21, no. 4 (2007): 313–319.

Piasecki, B. W. and G. A. Davis. "Restructuring Toxic Waste Controls: Intrinsic Difficulties and Historical Trends." pp. 1–13 in *America's Future in Toxic Waste Management: Lessons from Europe*, edited by B. W. Piasecki and G. Davis. New York: Quorum Books, 1987.

Picou, J. S. "The Talking Circle as Sociological Practice: Cultural Transformation of Chronic Disaster Impacts." *Sociological Practice: A Journal of Clinical and Applied Sociology* 2, no. 2 (2000): 77–97.

Picou, Steven. J. and Gill, Duane A. "The Exxon Valdez Oil Spill and Chronic Psychological Stress." *American Fisheries Symposium* 18 (1996): 879–893.

——— "Technological Disaster and Chronic Community Stress." *Society and Natural Resources* 11, no. 8 (1998): 795–815.

Picou, Steven J., Brent K. Marshall, and Duane A. Gill. "Disaster, Litigation, and the Corrosive Community." *Social Forces* 82, no. 4 (2004): 1493–1522.

Pinderhughes, R. "The Impact of Race on Environmental Quality: An Empirical and Theoretical Discussion." *Sociological Perspectives* 39, no. 2 (1996): 231–248.

Pinto, V. N. "E-Waste Hazard: The Impending Challenge." *Indian Journal of Occupational and Environmental Medicine* 12, no. 2 (2008): 65–70.

Plater, Z. J. B., R. H. Abrams, and W. Goldfarb. *Environmental Law and Policy: A Coursebook on Nature, Law, and Society.* St. Paul, MN: West Publishing Company, 1992.

Puckett, J., S. Westervelt, R. Gutierrez, and Y. Takamiya. *The Digital Dump: Exporting Re-Use and Abuse to Africa.* Seattle, WA: BAN, 2005.

Puckett, J., L. Byster, S. Westervelt, R. Gutierrez, S. Davis, A. Hussain, and M. Dutta. *Exporting Harm: The High-Tech Trashing of Asia. The Basel Action Network and Silicon Valley Toxics Coalition.* Seattle, WA: BAN, 2002. http://www.ban.org/E-Waste/technotrashfinalcomp.pdf (accessed December 10, 2009).

Quarantelli, E. L. and R. R. Dynes. "Response to Social Crisis and Disaster." *Annual Review of Sociology* 3 (1977): 23–49.

Raffensperger, Carolyn. "Precautionary Principle: Bearing Witness to and Alleviating Suffering, Part 1." *Rachel's Environment and Health News* 761 (March 6, 2003): 1–7. http://www.rachel.org

Raines, B. "Tests Reveal High Mercury in Some Gulf Fish." *The Times Picayune,* July 28, 2001, A-13.

Reich, Michael R. *Toxic Politics: Responding to Chemical Disasters.* Ithaca, NY: Cornell University Press, 1991.

Rao, P. and S. Sundar. "Statistical Issues in Accessing the Toxic Effects of Bhopal Gas Disaster." *International Statistical Review* 61, no. 2 (1993): 223–229.

Rhodes, E. L. *Environmental Justice in America: A New Paradigm.* Bloomington: Indiana University Press, 2003.

Rice, D. and E. Silbergeld. "Lead Neurotoxicity: Concordance of Human and Animal Research." pp. 659–675 in *Toxicology of Metals,* edited by L. W. Chang. New York: CRC-Lewis Publishers, 1996.

Robinson, Brett H. "E-Waste: An Assessment of Global Production and Environmental Impacts." *Science of the Total Environment* 408 (2009): 183–191.

Rosencranz, A. "Bhopal, Transnational Corporations, and Hazardous Technologies." *Ambio* 17, no. 5 (1988): 336–341.

Rosenthal, Erika. "The DBCP Pesticide Cases: Seeking Access to Justice to Make Agribusiness Accountable in the Global Economy." pp. 176–199 in *Agribusiness and Society: Corporate Responses to Environmentalism, Market Opportunities and Public Regulation,* edited by K. Jansen, and S. Vellema. New York, NY: Zed Books, 2004.

Saling, James H. and Andeen W. Fentiman. *Radioactive Waste Management,* 2nd ed. Boca Raton, FL: CRC Press, 2001.

Samdahl, D. M. and R. Robertson. "Social Determinants of Environmental Concern: Specification and Test of the Model." *Environment and Behavior* 21, no. 1 (1989): 57–81.

Sanchez-Murphy, L. "Economic Development and Environmental Threats: Tipping the Balance in Venezuela." *Loyola University Chicago International Law Review* 7, no. 1 (2009): 73–91.

Sarokin, D. and J. Schulkin. "Environmental Justice: Co-Evolution of Environmental Concerns and Social Justice." *Environmentalist* 14, no. 2 (1994): 121–129.

Satarug, S. and M. R. Moore. "Adverse Health Effects of Chronic Exposure to Low-level Cadmium in Foodstuffs and Cigarette Smoke." *Environmental Health Perspectives* 112, no. 10 (2004): 1099–1103.

Schafer, K. S. "One More Failed US Environmental Policy." *Foreign Policy in Focus* (May 31, 2006): 1–6. http://www.fpif.org/fpif txt/3492 (accessed August 2, 2010)

Schafer, K. S., S. E. Kegley, and S. Patton. *Nowhere to Hide: Persistent Toxic Chemicals in the US Food Supply.* San Francisco, CA: Pesticide Action Network and Commonweal, 2001.

Schaffer, Mathew. "Waste Lands: The Threat of Toxic Fertilizer." 2001. http://www.pirg.org/toxics/reports/wastelands/wastelands.pdf (accessed December 13, 2008).

Schecter, A., J. Startin, C. Wright, M. Kelly, O. Papke, A. Lis, M. Ball, and J. R. Olsen. "Congener-Specific Levels of Dioxins and Dibenzofurans in US Food and Estimated Daily Dioxin Toxic Equivalent Intake." *Environmental Health Perspectives* 102 (1994): 962–966.

Schierow, L. "The Emergency Planning and Community Right to Know Act: A Summary." *Congressional Research Service Report #RL 32683.* 2009. http//:fas.org/sgp/crs/misc/RL32683.pdf (accessed July 31, 2010).

Schmidt, C. W. 2006. "Unfair Trade: E-Waste in Africa" *Environmental Health Perspectives* 114, no. 4 (2006): A232–235.

———. "E-Junk Explosion." *Environmental Health Perspectives* 110, no. 4 (2002): A188–194.

Schnaiberg, Alan and Kenneth A. Gould. *Environment and Society: The Enduring Conflict.* New York, NY: St. Martin's Press, 1994.

Seki, Reiko. "Participatory Research by Niigata Minamata Disease Victims, and Empowerment of These Victims." *International Journal of Japanese Sociology* 15, no. 1 (2006): 26–39.

Sende, M. "Toxic Terrorism: A Crisis in Global Waste Trading." *ANAMESA* 8, no. 1 (2010): 30–41.

Sengupta, S. "25 Years Later, Toxic Waste Plagues Bhopal." *Seattle Times,* Monday, July 7, 2008. http://seattletimes.nwsource.com/cgi-bin /PrintStory.pl? (accessed November 22, 2009).

———. "Decades Later, Toxic Sludge Torments Bhopal." *New York Times,* July 7, 2008. http://www.nytimes.com/2008/07/07/world /asia/07bhopal.html (accessed November 22, 2009).

Setterberg, Fred and L. Shavelson. *Toxic Nation: The Fight to Save Our Communities from Chemical Contamination.* New York, NY: John Wiley and Sons, 1993.

Shabecoff, P. *Earth Rising: American Environmentalism in the 21st Century.* Covelo, CA: Island Press, 2000.

Sharma, D. "Bhopal: 20 Years On." *Lancet* 365 (January 2005): 111–112.

Sheppard, J. A. C. "The Black-White Environmental Concern Gap: An Examination of Environmental Paradigms." *Journal of Environmental Education* 26, no. 2 (1995): 24–35.

Short, James, F. "The Social Fabric at Risk: Toward the Social Transformation of Risk Analysis," *American Sociological Review* 49 (1984): 711–725.

Shrader-Frechette, Karen S. *Burying Uncertainty: Risk and the Case against Geological Disposal of Nuclear Waste.* Berkeley: University of California Press, 1993.

Shutkin, William A. *The Land That Could Be: Environmentalism and Democracy in the Twenty-First Century.* Cambridge, MA: MIT Press, 2000.

Signorini, S., P. M. Gerthoux, C. Dassi, M. Cazzaniga, P. Brambilla, N. Vincoli, and P. Mocarelli. "Environmental Exposure to Dioxin: The Seveso Experience." *Andrologia* 32 (2000): 263–270.

Simpson, Amelia. "Warren County's Legacy for Mexico's Border Maquiladoras." *Golden Gate University Environmental Law Journal* 1, no. 11 (2007): 153–174.

Singh, M. P. and S. Ghosh. "Bhopal Gas Tragedy: Model Simulation of the Dispersion, Scenario." *Journal of Hazardous Materials* 17 (1987): 1–22.

Smith, Eugene W. and Aileen M. Smith. *Minamata.* New York: Holt Rinehart and Winston, 1975.

Smith, K. R. "Environment and Health: Issues for the New US Administration." *Environment* 43, no. 4 (2001): 35–40.

Solomon, G. and T. Schettler. *Generations at Risk: Reproductive Health and the Environment.* Cambridge: MIT Press, 1999.

Sriramachari, S. "The Bhopal Gas Tragedy: An Environmental Disaster." *Current Science* 86, no. 7 (2004): 905-920.

SRNS. "Facts about the Savanna River Site: Transuranic Waste Program." 2008. http://www.srs.gov/general/news/factsheets/truw.pdf (accessed January 23, 2010).

Staggenborg, S. *Social Movements.* New York: Oxford University Press, 2011.

Steingraber, S. "The Social Production of Cancer: A Walk Upstream." pp. 19–38 in *Reclaiming the Environmental Debate: The Politics of Health in a Toxic Culture,* edited by R. Hofrichter. Cambridge: MIT Press, 2000.

Stringer, R., I. Labunska, K. Brigden, and D. Santillo. "Chemical Stockpiles at Union Carbide India Limited in Bhopal: An Investigation." Greenpeace Research Laboratories, Exeter, UK: Greenpeace. 2002. http://www.greenpeace.to/publications/bhopal%20stockpiles%20 report.pdf (November 24, 2009).

Strydom, Piet. *Risk, Environment and Society.* Philadelphia, PA: Open University Press, 2002.

Sullivan, K. "A Toxic Legacy on the Mexican Border: Abandoned US-Owned Smelter in Tijuana Blamed for Birth Defects, Health Ailments." *Washington Post* (February 16, 2003): A17.

Swan, Shanna, Eric Elkin, and Laura Fenster. "Have Sperm Densities Declined? A Reanalysis of Global Trend Data." *Environmental Health Perspectives* 105, no. 11 (1997): 1228–1232.

Swan, Shanna; Eric Elkin, and Laura Fenster. "The Question of Declining Sperm Density Revisited: An Analysis of 101 Studies Published 1934–1996." *Environmental Health Perspectives* 108, no. 10 (2000): 961–966.

Szasz, Andrew. *Ecopopulism: Toxic Waste and the Movement for Environmental Justice.* Minneapolis: University of Minnesota Press, 1994.

Sze, J. and J. K. London. "Environmental Justice at the Crossroads." *Sociology Compass* 2, no. 4 (2008): 1331–1354.

Tammemagi, Hans. *The Waste Crisis: Landfills, Incinerators, and the Search for a Sustainable Future.* New York, NY: Oxford University Press, 1999.

Taylor, Dorceta. E. "Blacks and the Environment: Toward an Explanation of the Concern Gap between Blacks and Whites." *Environment and Behavior* 21 (1989): 175–205.

——— "The Rise of the Environmental Justice Paradigm: Injustice Framing and the Social Construction of Environmental Discourses." *American Behavioral Scientist* 43, no. 4 (2000): 508–580.

Tesh, Sylvia N. *Uncertain Hazards: Environmental Activists and Scientific Proof.* Ithaca, NY: Cornell University Press, 2000.

The Chernobyl Forum. "Chernobyl's Legacy: Health, Environmental and Socio-Economic Impacts and Recommendations to the Governments of Belarus, the Russian Federation and Ukraine." 2003–2005. http:// www.iaea.org/publications/Booklets/Chernobyl/Chernobyl.pdf (accessed January 23, 2010).

The World Bank and CIDA. *Persistent Organic Pollutants and the Stockholm Convention: A Resource Guide.* Washington, DC: CIDA, 2001.

The World Resources Institute (WRI), United Nations Environmental Program (UNEP), United Nations Development Program (UNDP), and World Bank. *World Resources: Environmental Change and Human Health*. New York: Oxford University Press, 1998.

Thiele, L. P. *Environmentalism for a New Millennium: The Challenge of Coevolution*. New York, NY: Oxford University Press, 1999.

Thomas, J. K., J. S. Kodamanchaly, and P. M. Harveson. "Toxic Chemical Wastes and the Coincidence of Carcinogenic Mortality in Texas." *Society and Natural Resources* 11 (1998): 845–865.

Thomas, J. K., B. Qin, D. A. Howell, and B. E. Richardson. "Environmental Hazards and Rates of Female Breast Cancer Mortality in Texas." *Sociological Spectrum* 21, no. 3 (2001): 237–245.

Thornton, J. *Pandora's Poison: Chlorine, Health, and a New Environmental Strategy*. Cambridge: MIT Press, 2000.

Tierney, K. J. "Toward a Critical Sociology of Risk." *Sociological Forum* 14, no. 2 (1999): 215–242.

Tomain, Joseph P. "Nuclear Waste Policy Act (1982)." Major Acts of Congress. Encyclopedia. Com, 2004. http://www.encyclopedia.com /topic/Nuclear_Waste.aspx (accessed January 10, 2010).

Trotter, R.C., S. G. Day, and A. E. Love. "Bhopal, India and Union Carbide: The Second Tragedy." *Journal of Business Ethics* 8, no. 6 (1989): 439–454.

Tsoukala, T. H. "Science, Socioenvironmental Inequality, and Childhood Lead Poisoning." *Society and Natural Resources* 11 (1998): 743–754.

Tsydenova, O. and M. Bengtsson. *Environmental and Human Health Risks Associated with the End-of-Life Treatment of Electrical and Electronic Equipment*. Kanagawa, Japan: Institute for Global Environmental Studies (IGES), 2009.

Tucker, P. *Report of the Expert Panel Workshop on the Psychological Responses to Hazardous Substances*. Atlanta, GA: ATSDR, 1995.

Umesi, N. O. and S. Onyia. "Disposal of E-Wastes In Nigeria: An Appraisal of Regulations and Current Practices." *International Journal of Sustainable Development and World Ecology* 15, no. 6 (2008): 565–573.

UNEP. Report of the Intergovernmental Negotiating Committee for an International Legally Binding Instrument for Implementing International Action on Certain Persistent Organic Pollutants on the Work of its Fifth Session, Geneva: UNEP/POPS/INC. 5–7, (December 26, 2000).

———. "Global Mercury Assessment." 2002. Geneva, Switzerland: *UNEP*. http://www.unep.org/GC/GG22/document/UNEP-GC22-INF3.pdf (accessed December 18, 2009).

———. "E-Waste. The Hidden Side of IT Equipment's Manufacturing and Use." *Environment Alert Bulletin*. 2005. http://grid.unep.ch/product

/publicain/download/ew-ewaste.en.pdf (accessed February 14, 2010).

———. "Basel Conference Addresses Electronic Wastes Challenge." November 27, 2006. http://www.unep.org/documents/multilingual/default.print.asp? Document ID 485 & Article ID 543 (accessed November 27, 2009).

———. "Inventory Assessment Manual." *E-Waste*, 1 (2007). http://www.uep^.or.jp/ietc/publications/spc/ewastemanual_vol1.pdf (accessed December 18, 2009).

United Church of Christ Commission for Racial Justice. "Toxic Wastes and Race in the United States: A National Report on the Racial and Socioeconomic Characteristics of Communities with Hazardous Waste Sites." New York: United Church of Christ. 1987.

US Census Bureau. "Population Profile of the United States." Washington, DC: US Census Bureau. 2000. http://www.census.gov (accessed November 10, 2009).

US Department of Energy (DOE)."Transuranic Waste Processing Facts Sheets." 2009. http://www.becteljacobs.com/pdf/factsheet/TRU_waste_fact_sheet.pdf (accessed Ja-nuary 23, 2010).

US Department of Interior (DOI), Bureau of Mines (BOM).*Mining and Quarrying Trends in the Metals and Industrial Minerals Industries, 1991.* Washington, DC: DOT/BOM, 1993.

US Environmental Protection Agency (EPA). *Wastes from the Extraction and Beneficiation of Metallic Ores, Phosphate Rock, Asbestos, Overburden from Uranium Mining, and Oil Shale.* Washington, DC: EPA, 1985.

———. "Estimating Exposure to Dioxin-Like Compounds." *Volume II: Properties, Sources, Occurrence and Background Exposures.* Washington, DC: US EPA Office of Research and Development, 1994.

_____. *Five Year Review Report: Third Five-Year Review Report for Wells G & H Superfund Site, Woburn, Middlesex County, Massachusetts.* Boston, MA: EPA Region 1, 2009. http://www.epa.gov/region 1/superfund/sites/wellsgh/457903.pdf (accessed June 21, 2010).

———. *Understanding the Hazardous Waste Rules: A Handbook for Small Business.* Washington, DC: EPA, 1996.

_____.*Cleanup Begins at Wells G & H, One Year After Landmark New England Settlement: Superfund at Work-Hazardous Waste Cleanup Efforts Nationwide*, EPA 520-F-92-015, Washington, DC: EPA, 1993.

———. *The Preliminary Biennial RCRA Hazardous Waste Report.* Washington, DC: EPA, 1997.

———. *Toxic Release Inventory Public Data Release.* Washington, DC: EPA, 1999.

————. *Toxic Release Inventory Public Data Release*, 1999. Washington, DC: EPA, 2000.

————. "Electronics: A New Opportunity for Waste Prevention, Reuse, and Recycling. Solid Waste and Emergency Response." 2001. http://www.epa.gov/epaoswer/elec_fs .pdf (accessed May 22, 2009).

————. "Control of Hazardous Air Pollutants from Mobile Sources; Proposed Rule," *Federal Register*, 71, no 60 (2006):15804–15852.

————. "Radiation, Risks and Realities: Understanding Radiation in Your Life, Your World." 2007. http://www.epa.gov/rpdweb00/docs/402-k-07-006.pdf (accessed June 11, 2009).

————. "Management of Electronic Waste in the United States: Approach Two." EPA530-R-07-004b. Washington, DC: EPA, 2007. http://www.epa.gov/epawaste/conserve/materials/ecycling/docs/app-2.pdf (accessed December 29, 2009).

————. "Radiation: Risks and Realities." EPA-402-K-07-006. Washington, DC: EPA. 2007. http://www.epa.gov/rpdweb00/docs/402-K-07-006.pdf (accessed January 22, 2010).

————. "Electronic Waste Management in the United States: Approach 1." EPA530-R-08-009. Washington, DC: USEPA. 2008. http://www.epa.gov/epawaste/conserve/materials/ecycling/docs/app. 1. pdf (accessed December 29, 2009).

US Food and Drug Administration (FDA). 2000. *Total Diet Study*. Washington, DC: FDA.

————. 1999. *Food and Drug Administration Pesticide Program: Residue Monitoring*. Washington, DC: FDA.

USGS "Mineral Commodity Summaries: Cadmium." 2006. http://minerals.usgs.gov/minerals.usgs.gov/minerals/pubs/commodity/cadmium/mcs-2006-cadmi.pdf (accessed December 7, 2009).

————. "Mineral Commodity Summaries: Cadmium." 2008. http://www.minerals.usg.gov/minerals/pubs/commodity/cadmium/MCS-2008-cadm.pdf (accessed December 07, 2009).

Uyeki, E. and L. J. Holland. "Diffusion of Pro-Environment Attitudes?" *American Behavioral Scientist* 43, no. 4 (2000): 646–662.

Van Liere, K. D. and R. E. Dunlap. "The Social Bases of Environmental Concern: A Review of Hypothesis, Explanations and Empirical Evidence." *Public Opinion Quarterly* 44 (1980): 181–197.

Vaughan, E. and B. Nordestam. "The Perception of Environmental Risks among Ethnically Diverse Groups." *Journal of Cross-Cultural Psychology* 22 (1991): 29–60.

Waalkes, Michael P. and R. R. Misra. "Cadmium Carcinogenicity and Genotoxicity." pp. 231–243 in *Toxicology of Metals*, edited by L. W. Chang. New York: CRC-Lewis Publishers, 1996.

Wang, X. and J. Tian. "Health Risks Related to Residential Exposure to Cadmium in Zhenhe County, China." *Archives of Environmental Health* 59, no. 6 (2004): 324–330.

Wang, Z. and T. G. Rossman. "The Carcinogenicity of Arsenic." pp. 221–229 in *Toxicity of Metals*, edited by L. W. Chang. New York, NY: CRC-Lewis Publishers, 1996.

Wania, Frank and Donald Mackay. "The Evolution of Mass Balance Models of Persistent Organic Pollutant Fate in the Environment." *Environmental Pollution* 100 (1999): 223–240.

———. "Tracking the Distribution of Persistent Organic Pollutants." *Environmental Science and Technology* 30, no. 9 (1996): 390–396.

Wargo, J. *Our Children's Toxic Legacy: How Science and Law Fail to Protect Us from Pesticides.* New Haven, CT: Yale University Press, 1996.

Weir, David and M. Schapiro. *Circle of Poison: Pesticides and People in a Hungry World.* San Francisco, CA: Institute of Food and Development Policy, 1981.

WHO. "Evaluation of Certain Food Additives and Contaminants (Thirty-third Report of the Joint FAO/WHO Expert Committee on Food Additives)." *WHO Technical Report Series No. 776.* Geneva: WHO, 1989.

———. "Evaluation of Certain Food Additives and Contaminants (Forty-first Report of the Joint AO/Who Expert Committee on Food Additives)." *WHO Technical Report Series No. 837.* Geneva: WHO, 1993.

Widawsky, L. "In My Backyard: How Enabling Hazardous Waste Trade to Developing Nations Can Improve the Basel Convention's Ability to Achieve Environmental Justice." *Environmental Law* 38, no. 2 (2008): 577–625.

Widmer, R., H. Oswald-Krapf, D. Sinha-Khefriwal, M. Schnellmann, and H. Boni. "Global Perspectives on E-Waste." *Environmental Impact Assessment Review* 25 (2005): 436–458.

Williams, Eric. "Environmental Impacts in the Production of Personal Computers." pp. 41–72 in *Computers and the Environment: Understanding and Managing Their Impacts*, edited by Ruediger Kuehr and Eric Williams. Boston, MA: Kluwer Academic Publishers and United Nations University, 2003.

Williams, Eric, R. Kahhat, B. Allenby, E. Kavazanjian, J. Kim, and M. Xu. "Environmental, Social, and Economic Implications of Global Re-use and Recycling of Personal Computers." *Environmental Science and Technology* 42, no. 15 (2008): 6446–6454.

Williams, D. M. and N. Homedes. "The Impact of Maquiladoras on Health and Health Policy Along the US-Mexico Border." *Journal of Public Health Policy* 22, no. 3 (2001): 320–337.

Wirth, D. A. "Trade Implications of the Basel Convention Amendment Banning North-South Trade in Hazardous Wastes." *Review of European Community and International Environmental Law (RECIEL)* 7, no. 3 (1998): 237–248.

Wong, M. H., S. C. Wu, W. J. Deng, X. Z. Yu, Q. Luo, A. O. W. Leung, C. S. C Wong, W. J. Luksemburg, and A. S. Wong. "Export of Toxic Chemicals: A Review of the Case of Uncontrolled Electronic-Waste Recycling." *Environmental Pollution* 149 (2007): 131–140.

Worcester, Robert. "Public Opinion and the Environment." pp. 160–173 in *Greening the Millennium? The New Politics of the Environment*, edited by M. Jacobs. Malden, MA: Blackwell Publishers, 1997.

World Wildlife Fund (WWF). *Persistent Organic Pollutants: Hand-Me-Down Poisons That Threaten Wildlife and People.* Washington, DC: WWF, 1999.

Yandle, T. and D. Burton. 1996. "Reexamining Environmental Justice: A Statistical Analysis of Historical Hazardous Waste Landfill Siting Patterns in Metropolitan Texas." *Social Science Quarterly* 77, no. 3 (1996): 477–492.

Yassi, A., T. Kjellstrom, T. deKok, and T. L. Guidotti. *Basic Environmental Health.* New York, NY: Oxford University Press, 2001.

Youngman, N. "Understanding Disaster Vulnerability: Floods and Hurricanes." pp. 176–190 in *Twenty Lessons in Environmental Sociology*, edited by K. A. Gould and T. L. Lewis. New York: Oxford University Press, 2009

Yu, X. Z., Y. Gao, S. C. Wu, H. B. Zhang, K .C. Cheung, and M. H. Wong. "Distribution of Polycyclic Aromatic Hydrocarbons in Soils at Guiyu Area of China, Affected by Recycling of Electronic Waste Using Primitive Technologies." *Chemosphere* 65 (2006): 1500–1509.

Zastrow, C. *Social Problems: Issues and Solutions.* Belmont, CA: Wadsworth, 2000.

Zavestoski, S. "The Struggles for Justice in Bhopal: A New/Old Breed of Transnational Social Movement." *Global Social Policy* 9, no. 3 (2009): 383–407.

Zimmerman, R. "Social Equity and Environmental Risk." *Risk Analysis* 13, no. 6 (1993): 649–666.

ABOUT THE AUTHOR

Francis OlajideAdeola is professor of sociology at the University of New Orleans. He received his BS and MS from Arizona State University and his PhD from Mississippi State University. He regularly teaches courses in environmental sociology, social change, sociology of development, social organization, statistics, and research methods. His areas of research include environmental inequality at various levels, comparative study of environmental risks and environmental attitudes, environmental movements, cross-national development, poverty, hazardous wastes and health issues, and disasters of various etiologies. He has conducted extensive fieldwork in applied rural agricultural development in Nigeria, his country of origin.

He has published a number of book chapters and several peer-reviewed articles in a number of journals including *American Behavioral Scientist, Armed Forces and Society, Environment and Behavior, Human Ecology Review, Journal of Community Development Society, Journal of Third World Studies, Race, Gender & Class, Society and Natural Resources,* and *Sociological Spectrum.*

INDEX

Abelsohn, A., 93, 100
Aboud, R., 5
Acury, T.A., 22
Acute toxicity, 26
Adeola, F.O., 14, 19, 21–22, 30,
    38, 88, 96, 101, 116, 117, 133,
    154, 159, 166, 173, 180
Africa, 62–64, 65, 67, 83, 154, 181
Agency for Toxic Substances and
    Disease Registry (ATSDR),
    19, 28, 31–33, 37–40, 74, 107,
    114, 149
Agent Orange, 129, 170
Agriculture Street, 11, 21, 46, 91,
    94, 115–117
Agriculture Street Landfill (ASL),
    115–117
Agyeman, J., 175
Albrecht, S.L., 20
Allchin, D., 40
Amey, R.G., 20
Amzal, B., 35
Andrews, A., 50, 157
Arsenic, 17, 25, 28, 31, 32
Asante-Duah, D.K., 25, 26
Asbestos, 32, 33
Asia, 62, 83
Atomic Energy Act, 48

Bamako Convention, 155
Basel Action Network (BAN), 63,
    65, 72, 75, 76, 178
Basel Ban, 154–157
Basel Ban Amendment, 157
Basel Convention, 30, 82, 83,
    154–158, 160

Baum, A., 5, 15, 18, 21–22
Beck, Eckardt C., 9, 110
Beck, Ulrich, 7, 8, 13, 14, 167
Becquerel, Antoine Henry, 50
Bengtsson, M., 72
Berman, E., 32, 36
Bernard, A., 34, 35
Bhopal, 4, 11, 21, 46, 103, 108,
    117–127, 178, 180
Bio-accumulation, 34, 35, 40, 45,
    79, 90
Bio-magnification, 34, 40, 45, 90
Biomedical wastes, 27
Black Environment Network
    (BEN), 176, 179
Blackman Jr., W.C., 14
Blum, Elizabeth, 109, 171
Bodeen, C., 77, 78
Border Industrialization Program
    (BIP), 135
Broughton, T., 124
Brown, L.O., 6
Brown, P., 5, 6, 7, 14–15, 17, 22,
    115, 158
Bryant, B., 165–166, 168, 173
Bullard, R.D., 6, 8, 117, 163, 165–
    166, 168–169, 172–173
Bush Administration, 47, 180, 181

Cadmium, 33–36
Cancer Alley, 6
Cancer Corridor, 6, 21, 167
Carruthers, D.V., 138
Carson, Rachel, 5, 14, 15, 87, 92,
    147, 160
Catton, W., 4

CDC, 36, 79
Center for Disease Control (CDC), 36, 79
Chapman, S., 32, 33, 34, 147
Chedrese, J.P., 34, 35, 36
Chernobyl, 4, 21, 52, 108, 143
Clapp, J., 170
Clapp, R., 113
Clean Air Act, 33, 144
Cohen, G., 15
Colburn, T., 88, 89, 92
Cole, L., 168, 170
Comprehensive Environmental Response, Compensation and Liability Act (CERCLA), 114, 116, 127, 144–145, 148–150
Comte, A., 47
Conference of Plenipotentiaries (COP), 155, 157
Contaminated Community, 103, 107, 108
Convention on the Transboundary Effects of Industrial Accidents, 158–160
Corrosive community, 19, 22
Corrosivity, 27
Crawford, M., 32
CRTs, 34, 62, 65
Cunningham, M.A., 17
Cunningham, W.P., 17
Curie, 48, 51
Cuthberson, B.H., 19

Davidar, D., 122
Deep Water Horizon, 108
Department of Energy (DOE), 48, 49
Department of Interior (DOI), 29, 30
Department of Transportation (DOT), 26, 28
Dhara, V.R., 120–122
Dichlorodiphenyltrichloroethane (DDT), 14, 44–45, 86, 89–90
Digital divide, 63, 81
Dioxin, 44, 109, 127–134

Disaster, 1, 4, 6, 15, 103, 107, 109, 110
Dominant Western Worldview (DWW), 3
Douglas, M., 165
Drabeck, G.A., 3, 7, 16
Drotman, P.D., 36–38
D'Silva, T., 121
Dunlap, R.E., 4
DVD Players, 61, 70
Dynes, R.R., 3

Eckley, N., 90, 92
Edelstein, M.R., 5, 8, 14, 16, 18, 24, 107, 165
Emergency Planning and Community Right-to-Know Act (EPCRA), 23, 110, 149–151
Environmental equity, 168
Environmental justice, 6, 9, 10, 18, 163, 165–181
Environmental Justice Movement, 10, 11, 151, 156, 165–181
Environmental Protection Agency (EPA), 9, 15, 16, 25–27, 29, 33, 34, 37, 42, 46, 50–52, 59, 61, 62, 64, 76, 81, 113, 115–117, 140, 143–145, 147–150, 174
Environmental racism, 116, 168, 171, 180
EPCRA, 23, 110, 149–151
Epidemiology, 87
Epstein, S., 6, 15, 16, 29, 38, 78, 82, 88, 96
Erikson, Kai, 9, 14, 16, 68, 80, 87, 95, 107
Ethylmercury, 42
European Union (EU), 34, 76, 81, 82–83, 96, 103, 143, 151, 152, 154, 158, 160
EU Directives on WEEE, 82, 83, 85, 151, 152
E-waste, 15, 21, 34, 57–60, 62–72, 75–85

E-waste dumping, 81
Executive Order 12898, 167, 175
Externalities, 21, 69
Exxon Valdez, 17, 21, 143

Faber, D., 166, 170,
    171, 181
Faupel, C., 20
Federal Disaster Area, 112
Field, B.C., 15, 17
Field, M.K., 12, 15
Figueroa, R., 175
Fitzpatrick, K., 165
Flury-Herard, A., 47
Food and Agricultural
    Organization (FAO), 153
Food and Drug Administration
    (FDA), 42, 93–94, 95
Forum non-conveniens, 123
Foster, S.R., 168–170
Freudenberg, N., 7, 14, 114
Freudenburg, W.R., 7
Fritz, C.E., 3
Funabashi, H., 40

Garkovich, L., 20
Geider, T., 20
General Agreement on Trade and
    Tariffs (GATT), 155
Genetically modified organisms
    (GMOs), 159
Gerrard, Michael B., 27, 48
Ghana, 65
Gibbs, Lois M., 110,
    111, 171
Giddens, A., 7
Gill, Duane, 7, 19, 21
Girdner, E.J., 6
Globalization, 8
Global North, 13, 22, 23, 26, 57,
    61, 64, 70, 81
Global South, 22, 26, 33, 57, 61,
    63, 70, 81
Goklany, Indur M., 96, 97
Gore, Al, 88, 161, 175
Gould, K.A., 6, 15, 69

Government Accounting (or
    Accountability) Office (GAO),
    68, 76, 81–82, 136, 173, 181
Grasshopper effect, 90
Greenpeace, 72, 82, 125–126,
    132–134, 170
Green Revolution, 118
Grossman, E., 52, 60,
    66, 71
Guiyu, China, 21, 57, 77–79, 180

Hallman, W.K., 19
Hamada, R., 33–40
Hannigan, J., 20
Harmonized Tariff Schedule
    (HTS), 65
Hazardous Waste, 1, 13, 14, 16, 17,
    25, 27, 39
HDTVs, 64
Helsinki Convention, 144, 158
Hernan, R.E., 22, 128, 129,
    131–132
High Level Radioactive Waste,
    46–47, 50, 53
Hoffman-La Rouche, 127
Hofrichter, R., 86, 87
Homedes, N., 136
Hooker Electro-Chemical
    Corporation, 109–113
Hurricane Katrina, 16,
    20, 108

ICJB, 179, 181
ICMESA, 127–133
Ihonvbere, J.O., 134
Iles, A., 67, 68, 69, 81
Industrial revolution, 13, 29
International Atomic Energy
    Agency (IAEA), 46
iPads, 61
iPods, 61
"Itai-Itai" disease, 35, 36
Ivory Coast (Côte d'Ivoire), 158

Jablonski, S.M., 50
Johnson, B.L., 13

Kahhat, R., 72
Kapoor, R., 122
Kenedy, D.D., 114, 115
Kocher, S., 5
Koko, Nigeria, 21, 103, 133–135,
　　154, 180
Kreps, G.A., 3
Krueger, Jonathan, 82, 157
Kubasek, N.K., 145, 149, 152
Kummer, K., 82, 155, 156

Labunska, I., 125, 126
LaGory, M., 165
Lagrega, M.D., 15, 17, 26
Lallas, P., 88, 92
La Pierre, D., 119, 120
Latin America, 67
Lave, L.B., 6, 14
LCDs, 58
Lead, 36–38
Leaking Underground Storage
　　Tanks (LUST), 71, 148
Lemons, H., 3
Less Developed Countries (LDCs),
　　30, 65, 68, 77, 78, 80, 81, 123,
　　154, 169
Levi, D.S., 5
Levin, A., 6, 8, 17, 22, 109,
　　110, 171
Lichtveld, M.Y., 11
Lifton, R.J., 132
Lipman, Z., 134, 141
Locally undesirable Land Uses
　　(LULUs), 163, 165, 166
Love, William T., 108
Love Canal, 4, 6, 8, 9, 11, 17, 21,
　　45–46, 108–114, 143, 148,
　　154, 157, 158
Low Level Radioactive Waste,
　　46, 48
Luhmann, N., 7
Lupton, D., 165

Makofske, W.J., 16
Manhattan Project, 109

Maquiladoras, 135–138
McGinn, A.P., 26, 45
McGurty, E., 172
Medvedev, Z., 5, 52
Mercury, 38–42
Metales y derivados, 138–140
Methyl-Isocyanate (MIC),
　　118–122, 139
Methyl-mercury, 38–42
Midnight dumping, 18, 114, 172
Mikkelsen, E.J., 115
Miller, G.T., 17, 110
Minamata, Japan, 17, 40–42, 166
Minamata disease, 40–42
Mitchell, J.K., 108, 173
Mohai, P., 153, 165, 168, 173,
　　174, 177
Moore's law, 64
Moro, J., 119, 120, 124
MOSOP, 180
Muller, Paul, 45
Multinational Corporations
　　(MNCs), 117, 118, 133,
　　135, 170
Myers, J.P., 87, 98

NAFTA, 137, 138, 139
National Academy of Science, 30
National Institute of Health (NIH),
　　42
National People of Color
　　Environmental Leadership
　　Summit, 178, 183
National Priority List (NPL), 8, 20,
　　37, 39, 51, 60, 115, 117, 149
Nawrot, T.S., 34
Nebel, B.J., 17, 28, 43, 88
NGOs, 55, 61, 66, 69, 81, 178
Niagara Falls, 8, 109
Niagara Falls Board of Education,
　　109
Nickel-cadmium (NiCd) batteries,
　　21, 34
Nigeria, 21, 65–67, 76, 81, 88, 103,
　　133–135, 167–169

Nigg, M., 19
Not in My Backyard (NIMBY), 21,
    30, 154
Novotny, P., 173
Nriagu, J.O., 29
Nuclear Waste, 46–52
Nuclear Waste Policy Act, 46
Nuclear Weapons Complex, 51

Obama Administration, 48
O'Connor, J.O., 12, 15, 16
OECD, 82
Oliver-Smith, A., 4
Olsson, I.M., 35
Onyia, S., 63
Organization of African Unity
    (OAU), 155
Original equipment manufacturers
    (OEM), 59–60
Osame, M., 38–40
OSHA, 150
Oxychem, 82, 83, 84, 113

Pellow, D.N., 68, 170
Perrow, C., 5
Persistent Organic Pollutants
    (POPs), 9, 11, 21, 25, 43–46,
    55, 71, 85–86, 88–96, 99–101
Pesticide Action Network, 178
Pharmaceutical wastes, 27
pH scale, 26, 27
Picou, J.S., 7, 18–21
Pinto, V.N., 59
Planned obsolescence, 64, 68
Plutonium239, 48
Pollution Prevention Act, 15
Polychlorinated biphenyls (PCBs),
    9, 14, 16, 25, 31, 44, 45, 70,
    89, 95, 114, 115, 134, 172
Pope, C., 6
Popular epidemiology, 171
Precautionary principle, 86,
    96–98, 160
President Jimmy Carter, 113
Prince William Sound, 21

Principle of reverse onus, 98
Principles of environmental justice
    (PEJ), 167, 174, 183
Puckett, J., 60, 63, 65, 66,
    72, 76, 77

Quandt, S.A., 22
Quarantelli, E.L., 3

Radioactive Waste, 47–53
Raffensperger, C., 96
Raines, B., 30
Reactive wastes, 26
Record of Decision (ROD), 113
Recycling, 67, 68, 71, 72, 76,
    77–79, 80–81
Reich, M., 9, 112–113, 115–116,
    130, 132, 133
Resource Conservation and
    Recovery Act (RCRA), 14, 16,
    17, 25, 144, 145, 146–149
Restriction of Hazardous
    Substances (RoHS), 82–83,
    140, 151, 152
Rice, D., 30
Risk, 1, 38
Risk Society, 7, 13
Robinson, B.H., 77, 84
Rosencranz, A., 118
Rosenthal, E., 22
Rotterdam Convention, 153–154

Saha, R., 173, 174, 177
Saigo, B.W., 4, 17
Schaffer, M., 27
Schapiro, M., 90
Schierow, L., 150
Schmidt, C.W., 63, 66, 71, 72
Schnaiberg, A., 69
Setterberg, F., 14
Seveso, Italy, 4, 103, 128, 129, 133
Shavelson, L., 14
Short, J.F., 3
Shrader-Frechette, K., 47, 51, 52
Silbergeld, E., 30

Silicon Valley Toxic Coalition (SVTC), 68, 71
Silverman, G.S., 145, 149, 152
Smith, A.M., 39–40
Smith, E.W., 39–40
Smith, J., 5
Smith, K.R., 92
Social capital, 19, 163
Social problem, 20, 22
Sociocultural systems, 4
Solid Waste Disposal Act, 146
Stockholm Convention on Persistent Organic Pollutants, 55, 95
Stringer, R., 125–126
Strydom, P., 85
Sunrise city, 113
Superfund, 42, 71, 114
Superfund Amendments and Reauthorization Act (SARA), 149
Superfund law, see CERCLA
Synthetic Organic Compounds, 43–46

2,3,7,8-Tetrachloro-dibenzo-p-dioxin (TCDD), see Dioxin
Tammemagi, Hans, 1, 17, 26, 47, 52
Taxonomy of Wastes, 1, 25
Taylor, D.E., 166
TCPs, see Trichloro-phenols
TED case studies, 121
Tesla, N., 108
Tetrabromo-bisphenol-A, 70
Therapeutic community, 19
Third World, 43, 72, 90, 170
Thomas, J.K., 22
Thornton, J., 6, 14, 15, 16, 43, 71, 86, 92, 93, 96
Three Mile Island, 4, 21
Tierney, K.J., 7, 8
Times Beach, 21
Total diet study, 94–95
Toxic cycle, 15

Toxic disasters, 18, 22
Toxic gumbo, 20
Toxic Release Inventory, 15, 95
Toxic Substances Control Act (TSCA), 145, 172
Toxic Waste, 1, 6, 9, 13–17, 18, 21, 22
Transuranic Waste, 48–50, 53
Treadmill of toxics, 15
Trichloroethane, 71, 137
Trichloroethylene, 70
Trichloro-phenols (TCPs), 95, 110, 112
Tsoukala, T.H., 22
Tsydenova, O., 72
Tucker, 19

Umesi, N.O., 63
UNDP, 30, 38
UNEP, 30, 38, 61, 64, 68, 69, 70–71, 141, 142
Union Carbide Corporation (UCC), 117–118, 122–127
Union Carbide India Limited (UCIL), 118–127
United Church of Christ Commission for Racial Justice, 173
Upton, A.C., 6, 14
USGS, 34, 35

Valley of the Drums, 6
Volatile Organic Compounds, 114

Wandersman, A., 19
Wang, X., 32
Wang, Z., 32
Warren County, NC, 178
Waste Electrical and Electronic Equipment (WEEE), 58, 82–83, 151–152
Waste Isolation Pilot Plant (WHIPP), 49–50
Weir, D., 68, 90

Whalen, Robert, 111
Wildavsky, W., 165, 166
Williams, D.M., 136
Williams, E., 72
Wirth, D.A., 157
Woburn, Massachusetts, 103, 114, 115
World Bank, 30, 38
World Disaster Report, 107
World War II, 14, 35, 69, 87, 143
Wright, R.T., 14, 28, 43

WWF, 88, 92, 93

Xenobiotics, 4, 6, 10, 16, 70, 85, 92

Yassi, A., 88
Youngman, N., 4
Yucca Mountain, 47

Zastrow, C., 20
Zavestoski, S., 178
Zimmerman, R., 174

CPSIA information can be obtained at www.ICGtesting.com
Printed in the USA
LVOW071115280413

331212LV00006B/248/P